Spark

Spark

How Genius Ignites, From Child Prodigies to Late Bloomers

Claudia Kalb

Washington, D.C.

Published by National Geographic Partners, LLC
1145 17th Street NW Washington, DC 20036

The text permissions for this book appear on pages 353–354.

Library of Congress Cataloging-in-Publication Data

Names: Kalb, Claudia, author. | National Geographic Partners (U.S.)
Title: Spark : how genius ignites, from child prodigies to late bloomers /
    Claudia Kalb.
Description: Washington, DC : National Geographic Partners, LLC, [2021] |
    Includes bibliographical references and index. | Summary: "A look at
    genius, through portraits of 13 iconic figures in fields ranging from
    music to medicine, to explore what leads certain people to reach their
    creative heights early in life, whereas others don't uncover their
    potential until their later years"-- Provided by publisher.
Identifiers: LCCN 2020043781 (print) | LCCN 2020043782 (ebook) | ISBN
    9781426220937 (hardcover) | ISBN 9781426220944 (ebook)
Subjects: LCSH: Creative ability. | Genius. | Personality. | Psychology.
Classification: LCC BF408 .K224 2021 (print) | LCC BF408 (ebook) | DDC
    153.3/5--dc23
LC record available at https://lccn.loc.gov/2020043781
LC ebook record available at https://lccn.loc.gov/2020043782

Since 1888, the National Geographic Society has funded more than 13,000 research, exploration, and preservation projects around the world. National Geographic Partners distributes
a portion of the funds it receives from your purchase to National Geographic Society to
support programs, including the conservation of animals and their habitats.

Get closer to National Geographic explorers and photographers, and connect with our global
community. Join us today at nationalgeographic.com/join

For rights or permissions inquiries, please contact National Geographic Books Subsidiary
Rights: bookrights@natgeo.com

Interior design: Melissa Farris

Illustrations: Allison Bruns

Printed in the United States of America

21/BVG/1

To my grandparents,
Bella, Max, Sue, and Sam,
and to those who came before them.
And to the prodigies, midlifers,
and late bloomers in all of us.

Contents

Introduction

When I was child, I spent a lot of time wondering when I would know what I wanted to be. I've loved to write for as long as I can remember, and I spent many afternoons skimming through the pages of my parents' *New Yorker* magazines before I could make sense of the stories. Words fluttered through my mind like confetti at a ticker-tape parade.

Still, my path to becoming a writer was circuitous, and I have long been fascinated by the journeys of others. What role do our personality traits play in the livelihoods we pursue? Are we born with talent or lured by passion? How do we discover the spark that fuels our souls? And how do we know when we've found it?

These questions, which have stoked the minds of philosophers and psychologists for centuries, inspired me to explore the trajectories of 13 iconic figures who left colossal footprints in a variety of fields— from art and music to medicine, business, and politics. In my research, I became especially intrigued by the time line of discovery, which shapes the arc of this narrative. What propels some individuals to reach extraordinary creative heights in the earliest years of life while others uncover their destiny decades later?

Spark

The chapters in this book, best read sequentially, are organized not by birth order, but by the age at which genius ignites. Child prodigies Pablo Picasso, Shirley Temple, and Yo-Yo Ma, whose talents earned them acclaim in early childhood, launch the book, followed by Bill Gates, Isaac Newton, and entrepreneur Sara Blakely, whose moments of inspiration span the ages between 13 and 27. Julia Child, Maya Angelou, and Alexander Fleming constitute what I call midlifers—people whose momentous experiences transpired in their 30s and 40s. The chapters close with late bloomers Eleanor Roosevelt, Peter Mark Roget of *Roget's Thesaurus* fame, and the painter Anna Mary Robertson Moses, also known as Grandma Moses, all of whom initiated their most enduring work in the final decades of their lives.

I studied these individuals through a journalistic lens, conducting interviews and reporting on-site where possible, and bolstered my knowledge by delving into letters, memoirs, and biographies. At Woolsthorpe Manor, north of London, I stood in Isaac Newton's bedroom, with its limestone walls, and took in the view of his famed apple tree out the window. I went inside Alexander Fleming's lab, where he discovered penicillin, and took Eleanor Roosevelt's favorite walk through the woods at Val-Kill, her home in upstate New York. I listened to Yo-Yo Ma perform Bach's Cello Suites live at the Washington National Cathedral and at Tanglewood, and heard Sara Blakely speak to a group of entrepreneurs about her journey to founding Spanx, her multimillion-dollar clothing and shapewear company. Where possible, I talked to family members of the historical individuals profiled in these pages. I drew on a conversation I had with Ma for a magazine feature, and I interviewed Gates and Blakely.

From the cradles of civilization to the 21st century—Aristotle and Sappho to Jane Goodall, Alvin Ailey, and Isabel Allende—great minds have changed the way we understand ourselves and the world we inhabit. Their momentous achievements have fostered energetic

Introduction

debates about the inner workings of the brain and human behavior, and about how we judge ingenuity in art, music, literature, science, medicine, and technology. What are the origins of genius? What makes a genius? What *is* a genius?

This is inherently complicated terrain. Entire books, many of which informed my research, have attempted to explain human virtuosity. In these pages, I interpret the term with some degree of latitude. In certain cases, an individual's contributions are so immense and enduring that no other term seems sufficient. In others, the word acts more as an adjective describing a breakthrough or piece of creative work: The innovation itself is genius and the person who came up with it deserves ample recognition because of it.

Philosophers have long pondered the origins of genius. Plato described inspiration as a gift from the gods and likened the poet and prophet to an "empty vessel filled with divine infusion," says historian Darrin McMahon, author of *Divine Fury: A History of Genius.* Aristotelian thinkers, by contrast, attributed brilliance at least in part to biology, theorizing that an overabundance of black bile—one of the four bodily humors proposed by Hippocrates—endowed eminent souls with superior powers. Phrenologists attempted to find genius in bumps on the head; craniometrists collected skulls, which they probed, measured, and weighed.

Child prodigies, whose talent appears early in life, raise an eternally debated question: Is genius born or made? In the 19th century, the polymath Francis Galton proposed that the trait was passed down through family bloodlines; he mapped the lineages of an array of European leaders in disparate fields—from Mozart and Haydn to Byron, Chaucer, Titus, and Napoleon—to make his case. In 1869, Galton published his findings in *Hereditary Genius,* a book that launched the "nature versus nurture" debate and also spurred the misbegotten field of eugenics.

A century later, the focus shifted again—this time from pedigree to sweat equity. In a consequential 1993 study, psychologist Anders Ericsson and colleagues reported that the difference between elite and amateur musicians correlated strongly with how much time they spent in intensive study, or what Ericsson called "deliberate practice." These findings spawned the popularized notion that 10,000 hours of serious practice can lead to expert performance—a formula that vastly over-simplified Ericsson's research but was nonetheless seized upon by self-help gurus promising greatness in anyone determined enough to seek it. The reality almost certainly lies in between: Early ability engenders enthusiasm and drive, which leads to dedication and a desire to excel—a combination of nature and nurture.

No one has discovered a single source of genius, and such a thing will almost certainly never be found. But genius captivates us because of the vast potential it reveals. How can we, biological organisms comprised of 37 trillion cells, map a pathway to the moon, compose symphonies, solve mathematical quandaries, write poetry, and design new wireless technologies, one more mind-boggling than the next?

The subject of genius does not define the individuals in these pages. Instead, I use it as a framework to explore their lives through its core features: intelligence, creativity, perseverance, and luck. Although the 13 figures featured here differ in numerous ways—the eras in which they were born, the privileges and traumas they endured, the livelihoods they pursued—they share some combination of these characteristics, which have fostered their achievements and serve as common threads tying the first page to the last.

⌒

Intelligence has long been measured by a standardized IQ test, which evaluates a person's ability to perform challenging cognitive tasks—

mastering word puzzles or using logic to solve mathematical problems—to determine a score for "general intelligence." But the brilliance exhibited by the individuals profiled here clearly reveals that intelligence flourishes in different realms and in different ways. Bill Gates showed a remarkable aptitude for math beginning in childhood, an ability to think in numbers that seemed to come naturally. Maya Angelou had a facility for crafting imagery through words. Julia Child knew how to connect to people; her husband, Paul, dubbed this ability *"la Juliafication des gens*—the Juliafication of people," according to Child's grandnephew Alex Prud'homme. "She could charm a polecat," Paul marveled.

Pioneering achievements in any field would be impossible without creativity—an amalgamation of curiosity, openness to new experiences, imagination, and inventiveness. Yo-Yo Ma says his early childhood as an immigrant forged his quest to explore the world, inspiring his desire to connect people and culture through music. Isaac Newton had the vision to connect comets and falling apples to create a single law of universal gravitation. "He brought order out of chaos," wrote physicist Paul R. Heyl. Pablo Picasso's ability to see beyond the outlines of a human face upended conventional art. "With Picasso, the rules get exploded," his grandson Bernard tells me. "He's a kind of anarchist, but a creative one, not a destroyer."

Every one of these individuals exhibited perseverance, the ability to withstand difficult circumstances and push through obstacles. Several lost parents early in life (Newton, Alexander Fleming, Eleanor Roosevelt, Peter Roget) or children of their own (Roosevelt's third baby died; Anna Moses lost five of her 10 children). And in many cases, the trials of life inspired their work. Roosevelt overcame shyness and lack of confidence to become a public speaker and one of the great humanitarians of her generation. During the first half of her life, Angelou struggled to find meaningful work and later described those years as "a past of rejection, of slammed doors and blind alleys, of dead-end streets and culs-de-sac." She lived with the traumas of racism and rape but was never defeated. Instead, she penned words of pain and glory.

The enigmatic circumstances of our lives—where we're born, when we're born, to whom we're born—are defined by some measure of luck. Parents, teachers, and mentors play a pivotal role and come in myriad forms. Sara Blakely turns to a motivational speaker to help guide and buoy her. Roosevelt, orphaned at the age of nine, found comfort and counsel in an educator who nurtured her self-reliance and confidence. Gates's parents supported him even when he decided to leave Harvard in his junior year to start Microsoft. "I knew it was safe to try new things," he tells me, "because they were there for me."

Timing and serendipity are often keys to success. Child discovered French cooking because her husband got a diplomatic job that landed them in Paris. Grandma Moses' unlikely success as an octogenarian folk artist depended on a collector happening upon her paintings at a drugstore in Hoosick Falls, New York. Shirley Temple's career as a child star coincided with America's desire for an antidote to the Great Depression. Alexander Fleming's discovery of penicillin might never have happened if he hadn't left his lab dishes out on his worktable during a summer vacation.

In my research, I discovered numerous fallacies about how creative triumph comes to be. The notion of lone genius mythologizes the journey to achievement and has been replaced with an understanding that collaboration, in one form or another, is vital to the pursuit of new ideas. Newton's revelations did not emerge solely from the depths of his own knowledge; he built on the foresight of scientists who came before him, including Copernicus, Galileo, and Descartes. Gates created Microsoft with Paul Allen and started his philanthropic foundation with his wife, Melinda. Angelou had an editor in Robert Loomis, who prodded her to write her first book and then encouraged and supported her through dozens more over their four-decade partnership. When he retired in 2011, she remarked, "I can't imagine trusting a manuscript in the hands of anyone else." Ma dedicates much of his musical performance to collaborations with other musicians; his 2020 album, *Not Our First Goat Rodeo,* unites him and his cello with instrumentalists who play fiddle, banjo, mandolin, guitar, and bass. They learn from one another.

There are misperceptions, too, about how talent plays out. The prodigies in these pages all exhibited an adeptness nurtured by parents, but they were also gripped by a fascination for their craft and worked hard to improve. Picasso never stopped experimenting, painting at all hours of the night. Ma was born to musical parents who recognized his ability, but he still had to sit at the cello and practice; his joy of playing kept him going. Temple had an endearing brightness and appeal that captivated her audience, but she knew what was required to excel. "The rules were simple," she later wrote. "Hang on, keep tap shoes at the ready, and don't spill milk in the car."

Life trajectories do not follow a standard pattern. Most child prodigies do not grow up to be geniuses, no matter how flawlessly they master a skill; often, their solitary focus on a subject leaves them detached from peers and unable to navigate stresses as they age. Still, some prodigies do excel later in life, sometimes by choosing entirely new professions. After Shirley Temple retired from acting in early adulthood, she got married, raised three children, and kicked off a diplomatic career at the age of 41 that lasted more than two decades—the case of a prodigy and midlifer all in one.

Spark

Many readers will be pleased to discover that, yes, there is still time. The middle decades of life, long reputed to be the crisis years, may serve as a reawakening—a period when inadequacy and self-doubt dissipate and confidence and happiness surge. For the midlifers in these pages, the accumulation of decades brought courage and wisdom. Julia Child capitalized on her middle-agedness to entice her fans. Alexander Fleming's years of scientific know-how prepared him to recognize the significance of his findings. "The 50s," Maya Angelou told Oprah Winfrey, "are everything you've been meaning to be."

For those who despair about aging and fear an inevitable slowdown, consider the experiences of Roosevelt, Roget, and Moses. The first lady spearheaded the passage of the Universal Declaration of Human Rights at the age of 64. Roget, trained as a doctor, launched his second career writing and revising his thesaurus in his 70s. Moses spent most of her early life working on a farm and raising children; she turned to painting as a grandmother seeking a sense of purpose. "I have been painting a long time," she told the *New York Times* in 1945, the year she turned 85. "I better try something else. Perhaps I shall start building houses."

In the years to come, we may well learn more about our life trajectories from a biological viewpoint. Neuroscientists are looking for clues into how the "aha moment" transpires at the cortical level and how traits like resilience surface. Investigations into how the brain ages are revealing evidence that our neurons are far more proliferative than previously thought, allowing for more learning and growth at later stages in life. These studies cannot change what gives us joy or who we want to be. But they may provide intriguing insights into what we are capable of doing.

My hope is that these profiles will not only illuminate the many junctures at which discovery can happen, but will also inspire those who are still searching for fulfillment. Some lives are linear while others zigzag; some embrace a childhood love, while others find fulfillment in retirement. Every figure in these pages proves to be wholly human, filled with his or her share of missed opportunities, defeats, insecurities, and bouts of unhappiness.

One thing we can learn from these individuals is that the traits we most admire in others do not always come easily and we may need to be intentional about choosing them for ourselves. To this end, Blakely tells me she sets out to be courageous. Ma and Angelou, who called herself "a contrived optimist," make a determined point of being positive. "I have to really work very hard to find that flare of a kitchen match in a hurricane and claim it, shelter it, praise it," Angelou once said. Genius need not be vaulted onto a pedestal.

As in my first book, *Andy Warhol Was a Hoarder: Inside the Minds of History's Great Personalities,* I came across unexpected links between individuals in these pages that I was not aware of when I began my research. Picasso and Newton were both so feeble at birth that they were feared dead. Temple and Angelou lived radically different but parallel lives: Both were born in April 1928 and died within three months of each other in 2014. Roosevelt visited Shirley Temple in Hollywood in 1938, and Temple picnicked on lamb chops, potato chips, and ice cream with the Roosevelts at Val-Kill, Eleanor's home in upstate New York, a few months later. Roosevelt and Grandma Moses met in Washington, D.C., in 1949, when both won achievement awards from the Women's National Press Club. Paul Child, Julia's husband, organized an exhibition of Moses' paintings in Paris in 1950. Angelou kept a copy of *Roget's Thesaurus* by her side when writing. Bill Gates and Yo-Yo Ma, born just weeks apart in October 1955, were both undergrads at Harvard in the early 1970s. These overlaps signal a connection I believe exists among all of us.

I completed this book in the spring and summer of 2020, as the coronavirus surged—a cataclysmic upheaval that will define the early

Spark

21st century. The pandemic connected me to the past in a way I never expected when I began researching the historical figures I profile. I could not imagine the challenges that come with living through the onslaught of an unbridled virus prior to its emergence. For me, and I trust for readers as well, this experience makes the lives of several individuals in these pages ever more timely and relevant. Fleming witnessed the pernicious effects of the 1918 flu among soldiers and grappled with what caused it; his research helped inform his discovery of penicillin. Eleanor Roosevelt's husband, Franklin Delano Roosevelt, became sick with the 1918 flu while on a trip to Europe that summer—a turn of events that would change the course of the couple's relationship. And, most notably, Isaac Newton fled University of Cambridge in 1665 because of the spread of bubonic plague. His 17th-century quarantine in rural England—where his musings about calculus, gravity, and optics surged—suddenly became more tangible three and a half centuries later.

I felt, too, as if history were playing out before me as the three contemporary individuals in this book stepped up to offer their expertise as the world veered off-kilter: Blakely awarded five million dollars in loans to women struggling to run small businesses during the pandemic; Ma began streaming live performances to offer solace through music; and by midsummer, the Bill & Melinda Gates Foundation had pledged more than $350 million to the global health response, including to find a vaccine.

These pages conclude with a tribute to Leonardo da Vinci, whom I wrote about in a *National Geographic* story published in May 2019, on the 500th anniversary of his death. My reporting took me from Vinci, the Tuscan village of Leonardo's birth, to Amboise, France, where he lived out the last months of his life. If there is an ultimate expression of genius, it is in the curiosity, wisdom, and knowledge that Leonardo left behind. The artist's observations and studies of the elements around him—shadow and light, plants and earth, water and sky, bird flight and human anatomy—provide us a road map to the life within and around us. Through Leonardo, we better understand our existence in the universe.

I am deeply hopeful that these pages will leave readers with a renewed intention to embrace their own genius and offer it compassionately to others. If we are wise navigators of our lives, perhaps we can entrust a legacy worthy of our children and grandchildren. As Eleanor Roosevelt wrote: "To leave the world richer—that is the ultimate success."

CHAPTER 1

Pablo Picasso

The Modern Artist (1881–1973)

I t's the morning before Christie's Impressionist and Modern Art Evening sale in New York City, and suddenly, there it is. Just past the auction house's entrance at Rockefeller Center, Picasso's vibrant geometric portrait "Femme accroupie" ("Jacqueline") jaunts down a hallway, carried by two art handlers dressed in black.

Painted in the South of France in October 1954, the canvas features Jacqueline Roque, Picasso's 27-year-old mistress and soon-to-be wife, her arms clasped around a patchwork skirt of green and purple triangles. The artist, then 72, painted "Femme accroupie" in a single day; it gushes with vigorous brushstrokes, thick pigment, rambunctious shapes, misaligned eyes, and inverted nose. Golden light rings Jacqueline's body. Even off the wall, the painting commands attention.

That evening, auctioneer Adrien Meyer will start the bidding at $12 million. It will quickly surge upward as two Christie's representatives duel it out in a telephone bidding war on behalf of their anonymous clients. His back straight, his head jutting forward like a jaguar eyeing a peccary, Meyer will pivot between the pair until one of them signals defeat. Finally, with a bang of his hammer, he'll announce the winning price: $32.5 million.

Astounding, but not surprising. Half a century after his death, Picasso continues to bewitch, confuse, entice, and provoke with his bursts of shapes and color on canvas. Over the nine decades of his life, the artist mined his muses and his razor-sharp memory for subject matter: the pigeons out his childhood window, a tender glance from a lover, a child's smirk. And he worked voraciously, reinventing his style at a rapid pace—first his "blue" and "rose" periods, the African period, cubism, surrealism—creating tens of thousands of paintings, sculptures, drawings, copperplate etchings, and ceramics.

Picasso's passion for art erupted early in life. He communicated through pictures before he could speak. His first word was "piz," short for *lápiz*, or "pencil"; he drew spirals in the shape of a sugar cake. Art propelled him throughout his life, even if he could never explain it himself. "Where do I get this power of creating and forming? I don't know," he later said. "I have only one thought: work. I paint just as I breathe."

Picasso, the prodigy, benefited from family know-how. Like Mozart, Pablo had a parent in the business: His father, José, was a painter, and his son's first teacher. Expeditions together to the *corrida*, the dusty bullring located just off the Mediterranean Sea in his childhood town of Málaga, Spain, provided Pablo with rich subject matter. His earliest surviving works date to 1890, the year he turned nine, and include "Le Picador," a bullfighter on horseback. Soon, his precocious skill would begin to surpass that of his father.

Picasso, the man, was messy. He loved life at the circus and death at the bullfights. He could be boisterous and reclusive, amorous and domineering. But from his earliest days as a young artist to his final years painting musketeers and matadors, his life seemed destined for artistic greatness, his journey to genius fixed as firmly as paint on canvas. All the elements were there: a family that cultivated his creative passion; intellectual curiosity and grit; clusters of innovative provocateurs; and the good fortune to be born at a time of innovation—when new ideas in science, literature, and music fueled his work, and the advent of mass media catapulted him to fame.

The arc of Picasso's life was not only prodigious, but also long. He was 25 when he shocked the art world with "Les Demoiselles d'Avignon," in his mid 50s when he created his enormous antiwar mural, "Guernica," and 74 when, one rainy January day in 1956, he told biographer Antonina Vallentin that "I am only just beginning." Unlike other creative geniuses who died young—Sylvia Plath at age 30, Mozart at 35, van Gogh at 37—Picasso lived to the age of 91.

Sitting on a chartreuse couch and surrounded by paintings in his living room in Geneva, Picasso's son Claude contemplates the impact of his father's work: "He went on to destroy everything we were accustomed to," he says, "and created a new vision for everyone."

⌒⌒

Like a Rioja Gran Reserva aged in an oak barrel, artistic talent ripened as it traveled down Picasso's family bloodline. The artist's paternal grandfather, Diego Ruiz Almoguera, liked to draw, as did at least two of his children—his namesake, Diego, who grew up to be a diplomat with a side hobby of sketching portraits, and Picasso's father, who became a painter and art teacher.

Far less is known about the creative aspirations of Picasso's maternal side of the family, though there may have been at least one painter in the lineage, according to Picasso's biographer Roland Penrose. Either way, the marriage of José Ruiz Blasco and Maria Picasso López in December 1880 gave rise to one of the most celebrated artists in history, a baby with a very long name: Pablo Diego José Francisco de Paula Juan Nepomuceno María de los Remedios Alarcón Herrera Crispín Crispiniano Santísima Trinidad Ruiz Picasso. (Early on, the artist made a decision about how he wanted to be known: Shedding his father's common Spanish surname, Ruiz, he adopted his mother's more unusual and memorable Picasso.)

Pablo's listless arrival, close to midnight on October 25, 1881, in Málaga, would soon become family legend. The midwife feared the infant was stillborn, but Pablo's uncle Salvador, a doctor, revived him with a puff of cigar smoke that provoked a "bellow of fury" from the baby, Picasso later recounted. His mother had prayed for a son, and the couple's child had the added acclaim of being the first male heir on either side of the family. "If she had given birth to the Messiah," wrote Picasso's friend and longtime biographer John Richardson, "the family could hardly have been more joyful."

We know little about Picasso's earliest scribbles, many of which were scratched into the dust in Plaza de la Merced outside his childhood home—ephemeral musings lost to the wind or the rush of children's feet. Family members recalled that he would draw for hours at a time, sometimes taking requests for specific renderings—his cousin Maria's favorite was a donkey—until he was too exhausted to continue.

Historically, genius is cultivated by parents and teachers who nurture the seeds of greatness. Pablo grew up in a family that not only supported, but also indulged him. Surrounding the boy was a bevy of women—his grandmother, aunts, cousins, two younger sisters, and of course, Maria. "His mother was gaga over him," Claude tells me. Pablo's father recognized his son's artistic abilities, encouraged his interest, and gave him drawing lessons. And his uncle Salvador, whose income outdid his brother José's, became his nephew's benefactor, supporting his early art education. "It was a commitment by the whole family," says Claude, who is the legal administrator of his father's artistic estate.

Picasso's father struggled to make an income as a painter. He had adequate but modest talent, and his repertoire was limited largely to still lifes and the sporadic landscape. Many of his paintings featured pigeons, a favorite subject he shared with Pablo, who would incorporate the birds into his work throughout his life. Despite his own shortcomings, however, José doggedly trained his son, which gave the young Pablo grounding in the basics of realistic drawing and shaped his early technique. "He did very well," says Claude, "and he was the best student his father ever had."

In 1891, when Pablo was 10 years old, his father moved the family north to the city of Corunna (now known as La Coruña) to take a teaching job at a local art school. José needed the income to support his family of five—Pablo's younger sisters, Lola and Conchita, were born in 1884 and 1887, respectively—but it was a rocky journey by sea and a difficult transition for a man who was used to being surrounded by relatives and friends. Málaga, on Spain's Costa del Sol, infused its residents with sunlight and warm Mediterranean breezes; Corunna, a port city on the northwestern coast, doused José and his family with oceanic fog and rain. "At the first sight of Corunna," biographer Penrose writes, "he decided that he hated the place."

Pablo, by contrast, had a far more positive experience than his father. The move transpired as he began to mature out of childhood and marked a decisive moment in his evolution as an artist. No longer surrounded by the people he knew, Pablo discovered a refreshing independence that allowed him to wander the city streets with friends and to explore the region's landmarks and geography, all of which became subjects for his artwork—the rocky shoreline, the fishing boats in the harbor, and the Tower of Hercules, an ancient Roman lighthouse.

Corunna also enabled Pablo to transition from homeschooled pupil to official art student. In addition to taking his father's drawing class at the local art academy, he had the opportunity to learn from other instructors as well, including Isidoro Modesto Brocos, a painter and sculptor who had worked in Paris and whom Picasso long admired. Within a few years, Pablo was sketching in charcoal, experimenting with watercolor, and creating portraits of family members and friends.

In a pencil sketch of his sister Conchita, completed when he was about 13, Pablo depicts a girl with an appealing round face draped in short waves of hair coming out of a bonnet. Pablo was especially fond of Conchita, who was six years younger; when she became sick with diphtheria, he was so distraught that he pledged to abandon his passion. "In a burst of adolescent piety," according to Richardson, "he vowed to God that he would never paint or draw again if Conchita's life was spared." But his plea—and his promise—evaporated in the ocean fog: His sister died in January 1895, soon after her seventh birthday.

Over the months that followed, Pablo immersed himself in his art, which revealed a growing depth and sophistication. In oil on canvas he experimented with light, shadow, color, and composition, creating a series of portraits that included his father, a local beggar, a young girl in a red dress, and his dog, Clipper. Several of these paintings appeared in local shop windows, earning the young artist his first newspaper review at the age of 13. A writer for *La Voz de Galicia* noted the superior quality of the paintings, given the artist's age, and predicted a *"porvenir brillante*—bright future."

Over the next few years, Pablo's work gained wider recognition—first in Barcelona, where the family relocated in mid-1895, and then in Madrid. Barcelona offered not only a new teaching position for José at La Llotja, the city's school of fine arts, but also a far more cosmopolitan setting for Pablo's studies. At this point, the destinies of father and son became clear: José, a tall man in his late 50s with a linear nose and reddish beard, would turn his attention to overseeing the direction of Pablo's training. This occurred just as Pablo, a short, compact, dark-haired teenager, would develop an aversion to traditional academic instruction and begin seeking stylistic autonomy.

In an interview decades later, Picasso opined on how conventional rules made a mockery of art. "A good painting—any painting—ought to bristle with razor blades," he told the French novelist André Malraux in his studio in the winter of 1944. "Most painters make themselves a little cake-mold; then they make cakes. Always the same cakes. And they're very pleased with themselves. A painter should never do what people expect of him. A painter's worst enemy is style."

At La Llotja, Pablo's talent stood out, even in advanced courses among classmates several years older. The institute offered its students sketching sessions with live models, which Pablo embraced. But he skipped some of the more formalized classes—a pattern he would repeat at the esteemed Royal Academy of San Fernando in Madrid, an opportunity arranged by his father and subsidized by his uncle Salvador when he was 16 years old.

In Madrid, Pablo and his teacher, Muñoz Degrain, were far from simpatico. Pablo's disparagement of dogmatic instruction grew, while Degrain, a traditionalist, chided his student's work—Pablo's sketch of the city's Retiro Park looked like "a poached egg," his teacher told him. In a letter to a friend, written in the early days of his eight-month stay in Madrid, Pablo wrote that the teachers "haven't a grain of common sense. They just go on and on, as I suspected they would, about the same old things: Velázquez for painting, Michelangelo for sculpture, etc., etc." Pablo spent much of his time at the Prado Museum instead, studying a stylistically diverse selection of the masters—not just Velázquez, but also El Greco, Titian, and Rubens.

These museum visits further infused his passion for art. Even when he came down with a bout of scarlet fever in Madrid, which confined him to the shabby boarding room he had rented, Pablo did what he had to: took note of his surroundings and documented them on paper. A drawing from this period in his sketchbook, dated May 21, 1898, shows the objects that were scattered on the floor of his room—a hat, a pair of shoes, a urinal, and the legs of an easel.

Art, says his son Claude, was "the only thing he was interested in. That's the only thing he *was*. He was an artist through and through."

⌐

Little is known about the origins of exceptional talent early in life because it is so staggeringly difficult to study. Prodigies are generally defined as children whose abilities soar to a professional level by as young as 10 and into early adolescence. But they are rare, making it difficult to gather adequate sample sizes for scientific investigation. For academics who rely on grants to do their work, there is the added hurdle of funding: Research dollars tend to go to the study of illness and treatment. Prodigies are perceived as having positive attributes that don't need to be fixed, says David Henry Feldman, a pioneering researcher who helped define the field.

And then there's the hodgepodge challenge: Prodigies emerge at varying ages in different specialties. Precocious musicians tend to make it to the concert stage sooner than gifted young mathematicians solve complex problems, for example. How do you begin to measure talent when prodigies across domains posess disparate skills? A seven-year-old chess whiz is not the same as a seven-year-old who dances like Anna Pavlova. As a result, the study of prodigies is "a tiny empirical field," says Feldman.

Art prodigies can be especially tricky to study—first, because their work is so variable in style, and second, because art is expressive, rather than rules-based. This makes assessments far more subjective and prone to bias. And there's an access problem, too: Underprivileged kids don't often get the chance to take art lessons. Parents who do have the resources, meanwhile, are more likely to have their children learn an instrument, perhaps because the benefits of musical training have been more widely studied. As a result, some art prodigies likely are never discovered, says Ellen Winner, a developmental psychologist and director of the Arts and Mind Lab at Boston College.

Winner, who took drawing classes as a child and studied painting at the Museum of Fine Arts in Boston after college, is one of a handful of researchers who has spent decades researching the seeds of precocity in children. In 1987, she spent two months studying art education in China, where she observed six-year-old students master drawings of shrimps, crabs, and fish by replicating the simplified illustrations published in textbooks and created by their teachers. Winner discovered that the rules-based skills the children learned were useful; this approach allowed them to draw detailed and realistic images of real-life objects they had never seen, namely a Western-style baby stroller.

But what accounts for creative talent that shows itself prior to any kind of training? After returning to Boston, Winner began to identify fundamental characteristics among art prodigies—children who show unusual realistic drawing skill early on. She and her collaborator, Jennifer Drake, a developmental psychologist at Brooklyn College in New York, found that while art prodigies did not stand out on verbal and quantitative tests, they did excel at "local processing"—the ability to envision the parts that make up a whole object or to identify small shapes embedded within larger ones.

These exceptional perceptual skills are part of what realistic drawing requires—noticing precisely what the eye sees, as if it were a camera, while keeping one's own personal expectations and beliefs from interfering with the image. As early as age two, art prodigies are able to identify and draw the contours of a shape, like the roundness of an apple, which allows them to create recognizable likenesses. And years before their peers, they learn to illustrate the illusion of depth and three-dimensionality by incorporating foreshortening and perspective into their drawings.

These qualities mesh with what we know about Picasso. His visual memory was so prominent throughout his life it was practically a personality trait. The artist claimed to remember what his parents were wearing just after an earthquake struck Málaga when he was three years old on Christmas Eve (his mother threw a scarf around her head; his father, a cape over his shoulders). Friends and family members routinely remarked on his power of recall. "He hardly ever says anything, and it's impossible to know what he thinks, and he soon turns away," wrote fellow painter and lover Fernande Olivier, who lived with Picasso when he was in his 20s. "But everything registers, and he never forgets what he has seen."

Picasso's drawings also exhibited the early realism that Winner and Drake found in the prodigies they studied. His father started him off with focused instruction, teaching him to sketch the intricacies of pigeon feathers and feet; he drew meticulous representations of plaster casts of an arm or a leg as part of his conventional instruction at art school before he left it behind. Rooted in the academic tradition, he was faithful to shape, lines, and shading in his preliminary drawings—an achievement the artist remarked on himself decades later: "The attention to detail, the precision in them frighten me."

Much attention has been paid to the question of whether talent is innate or learned, an ongoing area of vigorous debate. On one end of the spectrum, psychologists influenced by the expansive work of Anders Ericsson argue that exceptional ability can be developed through intensive training and practice. In a landmark study of elite violin students in Germany that influenced the field, Ericsson found that the best musicians practiced more hours than their highly accomplished peers—7,410 hours by the time they were 18 years old, compared with 5,301 hours. "We found no shortcuts and no 'prodigies' who reached an expert level with relatively little practice," Ericsson wrote.

For those in this camp, "prodigy" is a misnomer, because it suggests that genes dictate behavior—and that therefore, children are born with talent. But those on the other side argue that *something* innate exists—if not a gene for music or tennis or art, then perhaps DNA that makes some people more persistent or predisposed to have a better memory or learn more quickly. In support of this theory, Zach Hambrick, a psychologist at Michigan State University, examined studies of chess players and musicians and found that practice accounted for only about a third of the difference in their skill level. Other factors, like starting young, were also important.

As with any complex human phenomenon, genius isn't a matter of nature versus nurture, but a combination of the two, with a wide range of characteristics playing a role. Personality traits like motivation and drive contribute to an individual's determination to succeed, as do the circumstances that nurture a child's passion.

For their part, Winner and Drake have found that talent sometimes shows up early in life, before it has been fully cultivated by an adult. They point to artwork by two-year-olds who haven't lived long enough to hone their skills through thousands of hours of repetition. "Anybody who's ever seen a prodigy knows it's totally implausible that it's due solely to deliberate practice," says Winner.

Based on their research, Winner and Drake believe that practice alone is not enough to achieve expertise. Talent, which they define as "the innate proclivity to learn easily and quickly in a particular domain," is a critical ingredient fueled by what Winner calls a "rage to master"—a passion so intense the children feel compelled to draw or paint whenever possible. Parents must pull them away from their sketch pads to sleep, eat breakfast, or get to school on time. "If you're really good," says Winner, "it motivates you to keep trying."

Picasso never doubted his own extraordinary talent. In his early 60s, he reflected on missing out on what he called "the genius of childhood"—the imaginative musings that disappear once maturity sets in—because he claimed he never experienced that innocence. "My very first drawings could never have appeared in an exhibition of children's drawings," he said with a mix of ego and remorse to the Hungarian-born French photographer known as Brassaï. "The child's awkwardness and naïveté were almost completely absent from them. I very quickly moved beyond the stage of that marvelous vision."

Few drawings from Picasso's first decade of life have survived; much of what we know about his early work comes from memories he shared with a close friend, Jaime Sabartès. As a result, "we have to make allowances for hyperbole or fantasy," biographer Richardson advises. This is true of anyone who resurrects childhood experiences—we all craft our own stories—but especially so among superstars whose reputations depend on wowing their audiences. Every letter, journal entry, memoir, and interview must be read with one overarching question in mind: Are these accounts inflated for a self-serving purpose?

That said, the drawings and paintings that Picasso had completed by his teenage years reveal a sophisticated eye for detail, composition, and mood that his educators and peers recognized—and that continue to impress experts today. His early work showed "huge maturity of feeling," says Suzanne Pagé, artistic director of the Fondation Louis Vuitton in Paris. "He was immediately very deep, very acute. He understood everything of the human world."

Pagé easily recalls how Picasso's singular talent for infusing his paintings with energy, sensuality, lust, and atmosphere "electrified" her the first time she saw his work in a gallery in Paris. "It was extra-ordinary," she says. "It was a bomb for me."

Indeed, prodigies often inspire powerful reactions. Many people are initially skeptical about their abilities, because they stretch so beyond the norm, says prodigy researcher Feldman. Are they really as they appear? How did they get that way? Why can't we explain it? The Latin *prodigium* carried the connotation of something that is "not just unexpected, but unwelcome and possibly dangerous and I think that notion persists in the public's mind," Feldman says. Watching a seven-year-old play Beethoven's piano sonatas while her peers are still learn-ing to jump rope can feel both awesome and unsettling at once. "It shakes your view of the world," he explains.

Picasso's grandson, Bernard Ruiz-Picasso, believes young Pablo may have inspired this mix of awe and fear. "His father might have been not only astonished," he says, "but petrified by the talent of his son."

⌒

In my quest to understand the origins of Picasso's artistry, I traveled to the sunlit town of Málaga, where landmarks of his childhood burst with vitality today. A choir sings *Man of La Mancha*'s "Impossible Dream" in the Church of Santiago, where Picasso was anointed with holy water as a baby. Plaza de la Merced, where the artist etched drawings into the dust, bustles with tourists sipping drinks at cafés and ordering a "hamburguesa Picasso." Pigeons light on the stones; the waters of the Alboran Sea lap at the shoreline; and Gypsies, who taught young Picasso to smoke through his nostrils and dance fla-menco, continue to traverse Málaga's streets.

In the courtyard of the Museo Picasso Málaga, I meet Bernard, who sips tea out of a red cup and reflects on how these early influences shaped his grandfather's mind and art. Everything about this place

Spark

is rich with history and sensuality, he says. Civilizations and religions—Phoenician, Roman, Jewish, Moorish, Christian, and Spanish—collided on the very soil Picasso inhabited. Aromas and pigments swirled in the air.

Gesturing to a nearby orange tree, Bernard says Picasso drew inspiration from the color of the fruits, from the violet flowers that drape Spain's jacaranda trees, and from the beige and white stones of Málaga's 11th-century Alcazaba, set into Gibralfaro hill just steps from the museum. "He kept in his mind all those senses, all those images, all those smells and colors, which nourished and enriched his brain," says Bernard, who established the Málaga museum with his mother, Christine Ruiz-Picasso, fulfilling his grandfather's wish.

Although this infusion of experiences surely imprinted Picasso in a personal way, the journey from prodigy to greatness is never a solitary pursuit. Creativity feeds on the exchange of ideas. Picasso found his first community of artistic comrades in Barcelona, where he returned after studying in Madrid. At El Quatre Gats café, he hobnobbed with more experienced Spanish artists who championed his work and helped organize his first noteworthy exhibition of portraiture. Together, these fellow creators contributed to "the stimulus that fueled the early stages of Picasso's rocket-like ascent," writes Richardson.

From an early age, Picasso set out to be an artist—he was *always* an artist—and in his teenage years he could finally take charge of his life trajectory. As much as he thrived in Barcelona, and as much as his close-knit family wanted him to stay, Picasso knew that his artistic growth depended on moving to Paris. He needed to branch out from his Andalusian upbringing and become proficient in a new language and unfamiliar surroundings; he needed to join an influx of young artists who would stoke his creativity and competitive drive. "He had to leave to confront the rest of the world and see whether he could stand up to the challenge," says Claude. "He was brave."

Pablo Picasso

Over the next several years, Picasso traveled back and forth from Spain to Paris before moving permanently in 1904, when he was 22 years old. In the Bateau Lavoir, a collective of budget artist studios in Paris's hilltop neighborhood of Montmartre, he lived with Fernande Olivier—the first of a long line of women whose bodies and lives became fodder for his art. A cluster of companions with exuberant minds surrounded him, including Guillaume Apollinaire, Gertrude Stein, Henri Matisse, André Derain, and Georges Braque, the man who would later become his partner in cubism.

Picasso's initial years in Paris provided the artist with both autonomy and the travails that come with it. He had little income to pay for food or furniture: His run-down studio was a mess, with its cast-iron stove held together by a twisted wire, and so cold in the winter that sips of tea left at the bottom of a cup would freeze overnight. Desperate for money, he begged for help from friends in Barcelona before finally landing his first devoted patrons.

Gertrude Stein and her brother Leo, early promoters of the artist's work, spent 800 francs on their initial visit to Picasso's studio—the first of numerous buying sprees that allowed them to amass a collection dominated by Picasso's "blue," "rose," and cubist periods, spanning 1901 to 1914. In a reminiscence decades later, Leo wrote an evocative description of meeting the 20-something artist: "One could not see Picasso without getting an indelible impression. His short, solid but somehow graceful figure, his firm head with the hair falling forward, careless but not slovenly, emphasized his extraordinary seeing eyes. I used to say that when Picasso had looked at a drawing or a print, I was surprised that anything was left on the paper, so absorbing was his gaze."

Prodigies are not destined to become geniuses, no matter how talented they are early in life. Genius requires a game-changing personality endowed with the courage and vision to transform a discipline.

Spark

Picasso was a boy when Paul Cézanne, Georges Seurat, and other post-Impressionists liberated themselves from the luminous brushwork of Impressionism, adding defined forms and emotional intensity to their canvases. When his turn came, Picasso seized his moment for change and charged forward with the intensity of a fighting bull. "He was so audacious you cannot even imagine," says Suzanne Pagé. "I suspect that even he did not know where he was going."

With his 1907 painting, "Les Demoiselles D'Avignon," the artist upended traditional composition, perspective, and aesthetic appeal. The canvas's depiction of five naked women at a brothel—their faces distorted, their bodies jagged—alarmed even Picasso's closest friends, including Braque. But the painting, completed when Picasso was 25 years old, would soon become the cornerstone of a radical art movement—cubism—and vault to the top of the list of most important paintings of the 20th century. In that moment, says Claude, "he brought down everything that anyone knew about art."

Picasso's fractured faces defined the cubist movement and shattered our most primal understanding of the world. Newborns are captivated by human faces, fixing their eyes on parents or caregivers in the very first days of life. By the age of four months—even before their visual systems are fully developed—babies are able to process faces almost as adeptly as adults, and they rely on those faces for information about how to understand the world. Picasso gave us mismatched eyes, elevated ears, and sideways lips, forcing us to rethink all that we knew to be true.

The artist was not alone in alienating viewers. Dean Keith Simonton, a longtime scholar of genius and professor emeritus of psychology at UC Davis, notes that Igor Stravinsky, Vaslav Nijinsky, and Marcel Duchamp—all contemporaries of Picasso—inspired uproars with their music, dance, and art, respectively. At the 1913 premiere of *The Rite of Spring* in Paris, which Stravinsky composed and Nijinsky choreographed and performed, audience members reacted to the atonal music and jerky dancing with loud catcalls and foot stamping. Duchamp's "Nude Descending a Staircase," painted in 1912, was demonized in the press—even by former president Theodore Roosevelt, who commented that the Navajo rug in his bathroom was a "far more satisfactory and decorative picture."

Creative genius is courageous and daring and forward-thinking in its very nature—otherwise, how would it represent change? Picasso and his fellow creators leaped into the unknown before others were ready to go. "They continued to take chances in pursuing their unique visions," Simonton writes in his book, *Greatness*. "Hence, the willingness to take creative risks often brings with it a special knack for alienating the public."

Picasso, who avoided commissions, never created art to please. "He painted whatever he wanted," Claude tells me, "and then he expected people to be interested."

So why do we find his work so compelling? I pose this question to José Lebrero Stals, artistic director of Museo Picasso Málaga. Picasso's talent, he believes, lies in his ability to mirror the very real anomalies in life through his art. When viewing his paintings, "people see some disharmony, what is not perfect," he says. And this is what we know to be true of the human condition: a combination of good and bad. In Picasso's flawed faces and bodies, we see ourselves.

An emerging field known as neuroaesthetics could help us understand how this interpretation might work at the cortical level. Using imaging technology, researchers are peering inside the brains of participants as they look at artworks. Their goal is to elucidate the relationship between viewer and painting: What happens when a person looks at a tranquil bouquet of flowers as compared with a battle scene filled with swords and storm clouds? How do we feel? And how do our brains respond?

Spark

In one study, Edward Vessel, a neuroscientist at the Max Planck Institute for Empirical Aesthetics in Frankfurt, scanned people's brains as they ranked their reactions to more than 100 images of artwork on a scale of 1 to 4, with 4 being the most highly moving. Unsurprisingly, the participants' visual system engaged every time they looked at a painting. But only the most moving artworks—the ones perceived to be especially beautiful or even striking or arresting—activated a series of networks in the brain known as the default mode network, which allows us to focus inward and access our most personal thoughts and feelings. Such a balance of outward viewing and inward contemplation is unusual, says Vessel. "It's a unique brain state."

This experience creates a special relationship between viewer and art, bringing the works alive in a meaningful way. Nobel laureate Eric Kandel is a neuroscientist and an avid art collector who owns two of Picasso's Vollard Suite etchings—a series of 100 prints Picasso created in the 1930s for the famed French art dealer Ambroise Vollard. Kandel says images that challenge, like Picasso's, recruit viewers into the creative process with the artist. Rather than act as bystanders, human brains are capable of taking incomplete clues and reconstructing fairly coherent images. "We have a tremendous ability to fill in details that are missing," he says.

But how? Kandel, co-director of Columbia University's Zuckerman Institute, and his collaborator, Daphna Shohamy, are hoping to find out by taking brain scans of participants as they complete a series of exercises with figurative and abstract paintings by Rothko, Piet Mondrian, Chuck Close, and other artists. Shohamy says they are eager to see whether abstract art elicits increased activity in the hippocampus, the brain's storehouse for memories. This would suggest, at a biological level, that humans intuitively draw on their own experiences when viewing and processing complex art.

Long before brain science could corroborate it, Picasso seems to have understood this dynamic. He rarely spoke about his work, allowing viewers to interpret it for themselves. "A picture lives a life like a living creature, undergoing the changes imposed on us by our life from day to day." he once said. "This is natural enough, as the picture lives only through the man who is looking at it."

Life journeys are often viewed through chronological age periods: childhood, midlife, and later years. One leads logically to the next. Picasso's trajectory was far more complex. As a child prodigy, he pushed forward, achieving artistic maturity early; as an adult he reached sideways and back, grabbing memories from wherever he could find them, defying time as he worked, reinventing his art and his relationships.

The artist had an extraordinary eye and an obsession for collecting, filling his disorderly rooms with piles of books and scatterings of glass or rocks. He found promise in everything, etching an owl or a goat onto a stone and then throwing it back into the sea. He formed the face of a sculpted baboon using two of his son's toy cars and crafted his famous "Bull's Head" out of a bicycle seat and rusty handlebars plucked from a junk pile. "The artist is a receptacle for emotions that come from all over the place," he once said, "from the sky, from the earth, from a scrap of paper, from a passing shape, from a spider's web."

Picasso's sharp and colossal memory served as a storehouse for inspiration. "He was a sponge," says Emilie Bouvard, a curator at the Picasso Museum in Paris. In her office, just steps away from the bustle of visitors at the museum, I ask Bouvard to pick the quality that best exemplifies Picasso's prowess. "In my opinion, it's *assemblage,*" she says, the artist's ability to assemble art from the multitudinous and

Spark

disparate memories in his mind—a conversation with a poet, the
haunting expressions in an El Greco painting, the medley of sensations
from Málaga, a pot of paint in his studio. As she reflects, Bouvard
calls up the French expression, *faire feu de tout bois,* "to make fire of
all wood." "That's the genius of Picasso," she says.

Picasso produced incessantly and in different mediums—paintings, collage, sculpture, lithography, ceramics, even jewelry. "He had
the ability to renew himself constantly," Diana Widmaier Picasso, an
art historian and curator and the granddaughter of Picasso and Marie-
Thérèse Walter, tells me. "He was so prolific, it's almost disarming."

Although Picasso had no idea where his creative bursts came from,
they rampaged through his head, disparate parts becoming whole
through his hands and his paintbrushes. "To know what you're going
to draw," he told Brassaï, "you have to begin drawing."

After early cubism, Picasso transfigured his work. Synthetic cubism, with the addition of collage, followed, then neoclassicism and
surrealism. He filled his canvases with harlequins, beach bathers, his
lovers, his children. For him, life and art were one. "He said he had
no secrets in his work," says Diana. "It was like a diary."

At the age of 55, Picasso issued one of art history's greatest indict-
ments of war. "Guernica," a 25-foot-long mural, was the artist's
response to an attack on the village of Guernica in April 1937 by
German and Italian bombers—an atrocity supported by Spain's fascist
leader Francisco Franco and nationalist rebels. In a palette of black,
grays, and white, he documented the hideous toll of war with mangled
bodies and the silent screams of Basque bombing victims in northern
Spain. Several years later, Picasso received a visit from a German officer
in Nazi-occupied Paris, who pointed to a photograph of his mural.
"Did you do that?" he asked. "No, you did," Picasso replied.

Even in his 60s, 70s, and 80s, Picasso was driven by an urgent
dedication to his art—a rage to master that never subsided. As his

biographer Vallentin noted, "With him, the surge of inspiration is not a gift of heaven patiently and humbly awaited, but a pressing need and a vital necessity."

As ideas cavorted through his mind, Picasso fought to bring them to life. He worked late at night and into the early morning, in his childhood and in his 90s. "It was almost neurological," says Diana, "something that forced him to be very active all the time."

He refused to stop until the work was flawless. "I can rarely keep myself from redoing a thing—umpteen times the same thing," he told Brassaï. "Sometimes it gets to be a real obsession. After all, why work otherwise, if not to better express the same thing? You must always seek perfection."

Talent, nurturing, opportunity, personality—Picasso had it all. He was also lucky. The artist came of age when photography overturned the need for traditional realism in paintings. The art world was primed for rule breaking and disruption, says András Szántó, a sociologist of art and museum consultant in New York City, and the media were newly equipped to celebrate it. Picasso, well aware of his stature, was masterful at branding his image. "He was so aware of his talent," says Olivier Widmaier Picasso, Diana's brother. "He understood that he would be important in the future, and people needed to understand the surroundings of his creations."

Early on, the artist became intentional about distinguishing himself and his influence. He took his mother's name, Picasso, not only because it was far more unusual than the common Spanish name Ruiz, but because he liked the elegant double *s*—evidence of his maternal great grandfather's Italian heritage and a distinctive sequence in the names of other great painters, including Matisse, Poussin, and Rousseau.

Picasso also thought ahead about the value of his work and the importance of leaving behind clues about his creative process, which he rarely discussed in public. Starting in his teens, he began methodically dating his paintings so they could be assembled in chronological order and analyzed by those who would study him later. It's not enough to know the work, Picasso once said, "One must also know when he made them, why, how, under what circumstances."

Savvy about the impact and reach of media, the artist often invited photographers to capture him posing with bravado in front of his canvases, dancing bare-chested with his lover, and playing with his children on the beach. In his lifetime, he vaulted onto three covers of *Time* magazine—first in 1939, when he was deemed "Art's Acrobat," and again in 1950 and 1980. Five years before he died, *Life* dedicated a 134-page double issue to his legacy. "He was able to layer his biography over these enormous inflection points in our culture," says Szántó. "He happened to play it really well."

The legacy of genius is a sweeping affair with eminence and acclaim often tied to personal anguish. The traits that promoted Picasso's creations—his infatuation with his work and his boldness—led to acclaim and even cultlike worship. Until Leonardo da Vinci's "Salvator Mundi" sold for more than $450 million in 2017, Picasso's $179 million "Les Femmes d'Alger" was the most expensive painting ever auctioned.

Picasso exhibits continue to draw record-breaking crowds; his works also serve as inspiration to collectors as far afield as Wang Zhongjun, a Chinese media mogul who periodically paints with a cigar clamped between his teeth and the Picasso he purchased in 2015, "Femme au Chignon Dans un Fauteuil," set up nearby.

But the creative insurgency that stoked Picasso's success also tainted his personal relationships, sometimes to the point of ruin. He craved women, and his charisma attracted them. Picasso had "a radiance, an inner fire," wrote his girlfriend Fernande Olivier, "and I couldn't resist this magnetism." But he was also jealous, egotistical, and misogynistic, displaying behaviors that are now fueling a public debate about whether an artist's conduct should affect the perception of his art.

At its most fundamental level, artistic genius shatters convention; one must be willing to alienate to experiment. But that same impulse can drive insupportable conduct. In 1891, Paul Gauguin, who would later influence Picasso, left his wife and five children when he traveled to Tahiti, where he painted his vibrant and evocative portraits of Polynesian women while fathering at least one child with a teenage bride. Jackson Pollock's explosive drip canvases catapulted him to acclaim as an abstract expressionist, but he was mercurial, self-absorbed, and unfaithful. His wife, the painter Lee Krasner, left him and fled to Europe, returning to New York only after learning that her 44-year-old husband had crashed his convertible while driving drunk in the Hamptons in August 1956. Only one passenger in Pollock's car survived: his lover, Ruth Kligman.

Are an artist's talent and his behavior inextricably linked? Not necessarily. Krasner's own work as an artist was overshadowed by her husband, but she has since gained recognition as a pioneer of abstract expressionism in her own right. An exhibition of her work at London's Barbican Art Gallery in 2019 impressed critics, one of whom described her energetic post-Pollock works as having a "percussive, Stravinsky-like rhythm."

Still, there may be a connection—at least among some artists—between ego and artistry. According to a study published in 2016, narcissism—determined in part by the size of an artist's signature—boosted the market price of an artist's work.

Picasso cycled through love affairs—many of them with women decades younger—perhaps to defy the odds of growing old. He married twice, first at the age of 36 and again at 79. In between, he had four children by three women. "It was not a simple family. It was a cubist family," his grandson Olivier tells me. "Everyone had the same creator, Picasso. But everyone has his own life."

Françoise Gilot, the mother of Claude Picasso and his sister, Paloma, met the artist in a Paris café in 1943 when she was 21 and he was 61. In *Life with Picasso,* published in 1964—and republished in 2019, just a few months before her 98th birthday—Gilot detailed Picasso's working and sleeping patterns (he often slept until noon); his superstitions (throw a hat on the bed and someone's going to die before the year is out); and his relationships with lovers and friends (Picasso took, they gave).

Over the course of their 10-year relationship, there were tender and passionate moments. Gilot, an accomplished artist in her own right, was enamored of Picasso the creator and Picasso the man. But she left him after a buildup of incidents that included him holding a cigarette against her cheek and threatening to throw her over the Pont Neuf and into the Seine. "The heart of the problem, I soon came to understand, was that with Pablo there must always be a victor and a vanquished," she wrote, and "with Pablo, the moment you were vanquished he lost all interest."

Tragedies piled up after the artist's death, with the suicides of Picasso's widow, Jacqueline Roque, his previous partner Marie-Thérèse Walter, and his grandson Pablito. In her 2001 memoir, Marina Picasso, Pablito's sister, describes the "Picasso virus" that poisoned her family: a plague of power, contempt, unkept promises, and lack of communication. "His brilliant oeuvre demanded human sacrifices," she writes. "He drove everyone who got near him to despair and engulfed them."

But other family members who never knew their grandfather, including Marina's half brother Bernard (he and Marina had different mothers but the same father—Picasso's son Paulo) and Olivier and

Pablo Picasso

Diana (their grandmother, Marie-Thérèse, had a 14-year affair with Picasso and was his most iconic model), have processed Picasso's life in disparate ways. While acknowledging the trauma, they also feel gratitude for Picasso's work and the fortune he left behind, which has not only deeply influenced the direction of their lives but also provided financial freedom.

All three are contributing to Picasso's legacy. Trained as a lawyer, Olivier has co-produced two documentaries and written several books about his grandfather "to set the record straight about rumors and legends and truths." When I meet him at his apartment in Paris, I can't help but notice that like Picasso's son Claude, he resembles the artist with his dark hair and eyes. I ask if he agrees, and he answers with a laugh, saying he did not inherit Picasso's talent. "Probably the nose, sometimes the eyes, but never the hands," he says. "Genius is not transferable."

Diana, who studied art history at the Sorbonne, specialized in drawings by the old masters at Sotheby's before deciding to dedicate herself to researching her grandfather's work. She has authored a book focusing on the eroticism that courses through the artist's work and has curated numerous exhibitions, including one at New York's Gagosian Gallery that focused on Picasso's paintings of her mother, Maya, the artist's eldest daughter. Now at work on what she anticipates will be a four-volume catalogue raisonné of the artist's sculptures, Diana says her grandfather left her with a personal mission to be tenacious and make sure she attends to her projects with the same thoroughness that he did. "We have an obligation," she tells me. "I'm always doubting that I got it right. I want to do it again. That's what makes me excited."

When I ask Bernard, who was 13 when Picasso died, what he remembers about his grandfather, he tells me he has warm memories of visiting him on the French Riviera in the summers, swimming and

Spark

gathering with family on the beach. (Bernard's mother was Christine Ruiz-Picasso, the second wife of Paulo, Picasso's oldest son.) "He was an old guy, but he had a strong presence," he says, as birdsong fills the air in Málaga. "He was really looking and paying attention to life." In addition to overseeing the Picasso Museum in Málaga, which opened in 2003, Bernard and his wife, Almine Rech-Picasso, established an art foundation—Fundación Almine y Bernard Ruiz-Picasso para el Arte—around his grandfather's work.

"Life is full of drama. We are not the only ones," Bernard tells me. "I'm deeply grateful for what Picasso gave me."

⌁

Although he was never granted French citizenship, Picasso lived his adult life in Paris and in the South of France, where he spent his summers and the last two decades of his life with his second wife, Jacqueline Roque. The French Riviera took the artist back to the warm waters of the Mediterranean and to the bullfights of his youth, which he now watched at arenas in Arles, Nîmes, and Vallauris.

In 1960, when he was 78 years old, Picasso produced 14 ink wash drawings (one of which featured colorful pastel) of bulls and picadors in a single day. The year he turned 88, he created an astounding 167 paintings and 50 drawings, which were exhibited at the Palace of the Popes in Avignon, France. The show's organizer told the *New York Times* that, like the great Renaissance artist Titian, "the older [Picasso] grew, the more his work exuded an impression of radiant youth."

In the fall of 1971, the Louvre opened its first exhibition of a living artist to celebrate Picasso's 90th birthday. In a tribute, France's president Georges Pompidou called him "a constantly boiling volcano."

A year and a half later, he was gone. On the evening of April 7, 1973, Picasso had dinner with friends, where he "was reported to have been the gayest member of the party, eating heartily and telling stories," according to a news account. At around 11:30 p.m., he headed back to his studio to work. He died the next morning, after suffering a pulmonary edema at his hilltop villa overlooking the sea in Mougins. He left behind some 15,000 paintings and drawings, several thousand engravings, and hundreds of sculptures and ceramics.

In the end, Picasso's journey from prodigy to legacy is a story of ultimate conquest. "Every time he went and attacked something, he did it in a very unconventional way," says Claude as he sits surrounded by both his father's and his mother's paintings in his home, the midday sun streaming in. "He left few corners untouched and unturned."

Still, when I ask how he explains his father's genius, he answers with the most uncomplicated reply: "How do I explain it? I don't explain it," he says. "I just understood it. It was obvious to me as a tiny child."

CHAPTER 2

Shirley Temple

The Child Star (1928–2014)

or Picasso, it started with his hands. From somewhere deep inside his being, a young boy's artistry emerged as he etched drawings into the dirt under the jacaranda trees in Málaga, Spain. Some 50 years later, at a dance school in Los Angeles, Shirley Temple launched her journey with her feet. At the age of three, while other children were still figuring out how to pedal a tricycle, Shirley was prancing, twirling, tapping, and learning how to fall gracefully.

Like Picasso, Shirley Temple had a drive to succeed and a knack for excelling at what she loved. Her facility for movement, her ear for rhythm, her physical dexterity, her ability to listen and learn—these were traits that emerged early and stood out.

At just four years old, Shirley strides across the stage with her signature curls and sparkles in the short 1933 film *Kid in Hollywood*. A pint-size boy playing a movie director hands her a bouquet of flowers and says, "You were wonderful! You'll be my new star!" Shirley, cast as Morelegs Sweettrick, a spoof of Marlene Dietrich, replies, "Oh, Mr. Director," at which point a quartet of tiny yes-men shout out, "A new discovery! Amazing! Stupendous! Colossal!"

And so she was. With her dimples and smile and toddler physique, Shirley's angelic appearance captivated viewers of all ages and her magnetic personality kept them watching. Beginning with her break-through roles, film critics from coast to coast gushed over her ability to command the screen. A *Los Angeles Times* review of the 1934 movie *Little Miss Marker* described her as "a new wonder child" while the *New York Times* effused that she "is virtually the stellar performer in the present work, and no more engaging child has been beheld on the screen."

By the age of six, Shirley had achieved box office stardom, bumping out experienced superstars, including Claudette Colbert and Clark Gable. In her two decades as an actress, she appeared in 43 feature films, inspired look-alike dolls and Shirley Temple wannabes, built a multimillion-person fan base that included a young Andy Warhol, and charmed the likes of Eleanor Roosevelt and J. Edgar Hoover.

Some people meander through life or, like Picasso, ascend and never come down. Temple grounded herself in an authenticity that allowed her to flourish in seemingly disparate phases of her life: child star at four; wife and mother in her 20s and 30s; international activist and diplomat in her 40s, 50s, and 60s. Her natural affability, unadorned sense of humor, tough inner core, and unflagging optimism buoyed her through a litany of experiences most people never have—whether a swarm of fans vying for her attention or Russian tanks invading Prague.

How did Shirley Temple avoid the fate of prodigies who burn out by early adulthood? What propelled her to transform her life, not just once but twice? Both nature and nurture came into play. Shirley had talents that her mother, fellow actors, and directors cultivated—but she also experienced life's hurdles at an exceedingly early age. Acting on stage as a toddler required stamina, determination, and sociability, all of which prepared her to navigate the drama of world affairs as a diplomat.

As a child actress beloved for her diminutive size and her youthful innocence, Shirley's precocious appeal had an expiration date—a reality she seemed to understand and accept early on. But even as she aged out of Hollywood, she mustered an optimistic outlook, rejecting sentimental ruminations over a past that could never be recaptured. "In some lives the pleasures of retrospect and testimonial laurels hanging on the wall are comforting," she wrote in her autobiography, *Child Star*. "For me, past achievement is dry as dust, static memories best elevated to attic storage." Her focus was always the next adventure—tomorrow's challenge, whether becoming a mother or running for Congress.

In 1977, when she was 49 years old, Temple received a Life Achievement Award from the American Center of Films for Children on the lot of 20th Century Fox. "Shirley has proved a child actor can be something besides having dimples and being able to cry on cue," fellow child star Darryl Hickman said in a tribute. In accepting her award, Temple made clear the conviction that drove her life, from beginning to end. "The most important moment," she said, "is now."

⌒⌒

Shirley Temple was a wish-come-true baby. Her parents, Gertrude and George Temple, met in 1910 at Kramer's dance hall in Los Angeles, got married, and had two sons: Jack and George, Jr. But when Gertrude was 34 years old and the boys were 12 and 8, she "announced her intention to produce a baby girl," Shirley later reflected. Neither Gertrude's age nor her sons' ages could deter her quest.

Spark

George did his part, even agreeing to have his tonsils removed—a common procedure at the time—after his doctor suggested it might increase the odds of having a girl. Once nature complied, Gertrude set out to enlighten her growing baby—playing classical music on the radio, reading books aloud, and meandering through museum exhibitions, where she purposefully absorbed the artwork on display. "It was her mystical, Teutonic conviction that noble thoughts, beautiful sights, and pleasant sounds could somehow imprint themselves directly on her child," Shirley reflected, "a prenatal blitzkrieg."

As a teenager, Gertrude had fantasized about becoming a ballerina, but her parents didn't think it a worthy endeavor and felt she should find herself a husband instead. The birth of Shirley, on April 23, 1928, gave Gertrude an opportunity to recapture her joy. She entertained her baby by prancing around the living room with the radio turned up.

When Shirley was barely three years old, Gertrude enrolled her at Meglin's Dance Studios in Los Angeles, where director Ethel Meglin taught her students myriad moves—from a single tap time step to tango sequences and balletic pliés. "Mother put me there, because I had so much energy that she thought I was going to tear the house down," Shirley later told Larry King, "so she said, 'Let's do something with this.'"

Over two and half years, Gertrude drove Shirley to Meglin's, and Meglin inspired her to dance. A former Ziegfeld Follies performer, Ethel Meglin put up with the unavoidable toddler pileups. But she and her assistants persisted in recurrent run-throughs until their students mastered their lessons. The repetition wasn't always fun, but it made every motion "as reflexive and natural as walking or standing," Shirley later wrote. "It could have been one big bore, had I not really enjoyed myself."

Energetic kids tend to hopscotch through life—hiding under pillows one minute, banging out music on a kitchen pot the next.

Shirley Temple

Few have the focus required to master a single subject. A rule of thumb among child experts is that children should be able to stay focused for two to five minutes multiplied by their age—which sounds about right to Melissa Tonnessen, a second-grade teacher in New Jersey. Many kids find it difficult to sit still, listen, and follow directions, she says, and this is to be expected. "I look at the students I teach every day," she says, "and I can't imagine any of them being able to do what Shirley Temple did."

Shirley had the ability to stay focused and the determination to learn, a quality that her daughter Susan believes cannot be taught. "She was preternaturally gifted," Susan tells me. "Most three-year-olds are not concerned with things beyond their immediate self." And her mother was intuitive, too. "She could read people from a very young age," says Susan.

Shirley's big break came when a movie producer walked into the studio looking for talent one day. The Meglin Kiddies, as they were known, were asked to line up and show off their best time step. Put off by the look of the producer's jowly face, Shirley hid under the piano, but she'd already been noticed. "As fate would have it, he said, 'I want that one,'" she later recalled. "That was me. That was my start."

Within weeks, the fledgling three-year-old dancer had completed a screen test and was signed by a production company called Educational Films Corporation to star in a series of 10-minute films known as *Baby Burlesks*. The bizarre shorts, for which Shirley earned $10 a day during filming (rehearsals went unpaid), featured kids wearing diapers with oversize safety pins as they impersonated Hollywood stars and satirized adult movies.

In her debut in *Runt Page* (a takeoff on *The Front Page*), a toddling Shirley plays a supporting actress amid a cast of costars, all of them bare chested. By film number two, *War Babies* (lampooning the World War I film *What Price Glory*), she'd earned the romantic lead as a

French bar girl, in which she sashays around stage ogled by feuding military boys who call her "baby" and woo her with a giant lollipop.

Decades later, Shirley provided an honest account of the *Baby Burlesks*. The hours were long, the costumes skimpy, and the punishment cruel. Disobedient kids were given the black box treatment—a time-out in a six-foot-square portable workstation intended for sound technicians. Misbehaving children, sweaty from performing under the lights, were ushered into the dark space, in which they had nothing to sit on but a large block of ice. "It was really a devilish punishment," Temple wrote.

Above all, the children were taught to memorize lines and mimic facial contortions and moves that were steeped in flirtatious behavior, sexual innuendo, and stereotyping. In *Polly Tix in Washington,* five-year-old Shirley plays a "strumpet," as she later described her role, hired to win over a Washington senator. In her boudoir, Polly (based loosely on Mae West) is attended by a maid dressed in an apron (played by a Black child uncredited in the film) before she enters the senator's office and saunters up to him, draped in pearls and bracelets. "Oh hello, I'm Polly Tix. Boss Flint Eye sent me over to entertain you," she says. "Although generally a heavy-handed spoof," Temple wrote, "the films were a cynical exploitation of our childish innocence, and occasionally were racist or sexist."

Still, in later interviews and in her memoir, Temple said she viewed the experience as an education that prepared her for the harsh reality of acting. During her two-year stint performing *Baby Burlesks,* starting in late 1931 and ending in 1933, she spent long days on set (her mother recalled one 11.5-hour Saturday) and quickly recognized that that the job required a can-do attitude, even when it meant kissing other stars—"including some unattractive ones," she noted.

Temple later credited her mother with teaching her to pose and gesture. But it was in the studio that she learned where to stand and how to hold her head by sensing the warmth from stage lights on her face—a skill she would rely on for the rest of her career. A quick study, Shirley figured out the most important lessons: Know your lines, don't waste time, and work hard if you want to act.

Before long, Gertrude was toting her daughter around Hollywood to feature-length movie auditions. "If there was a door open," Shirley later said, "we'd go through it." She didn't get every part—MGM turned her down to play a child version of Joan Crawford in *Dancing Lady*, which also featured newcomers Fred Astaire and Nelson Eddy— but the cameos started stacking up. In December 1933, at the height of the Great Depression, Shirley wowed the execs at Fox Studios during filming for *Stand Up and Cheer!*, a feel-good movie whose premise was to buoy a weary nation.

In its winning number, "Baby, Take a Bow," Shirley sings and tap-dances in a starched polka-dotted dress in flawless synchrony with actor James Dunn. It didn't matter that she had tripped just before filming and cut her forehead—her mother took care of that by pasting down a curl to cover the gash. She stole the show and earned a contract with Fox for $150 a week with an option for seven years.

On December 21, 1933, Shirley Temple signed the paperwork with a backward "S."

⌁

Timing played a key role in Shirley Temple's early success: She was adorable and wholesome and uplifting at a time when the country desperately needed an emotional boost. Some 13 million Americans, a quarter of the workforce, were unemployed; factories had been shuttered; and wages, for those lucky enough to find a job, had plummeted. Temple's early career also coincided with the advent of talking motion pictures, which debuted in the late 1920s. Viewers

Spark

got to take in every bit of their beloved star—not only her looks, but the sound of her voice as well. "It is a splendid thing that for just 15 cents, an American can go to a movie and look at the smiling face of a baby and forget his troubles," President Franklin D. Roosevelt famously said.

Even when critics panned a film, they singled out Shirley, who was almost always cast as the cheerer-upper, the little girl who could mend relationships and make everyone happy. Writing about *Baby, Take a Bow* in June 1934, *Los Angeles Times* drama editor Edwin Schallert admitted that he was skeptical about child wonders. "However, this little girl Shirley," he wrote, "is very real and very different." The *New York Times,* meanwhile, declared that Shirley "tucks the picture under her little arm and toddles off with it."

In 1934—two years after her first on-screen appearance—Fox created *Bright Eyes,* the first film written specifically for their new child wonder. The movie featured Shirley singing "On the Good Ship Lollipop" in a checkered dress with a bow fastened to the collar. The song was an instant hit, selling 500,000 copies of sheet music. In that breakthrough year alone, she appeared in nine movies, and in the years to come dozens more, including *The Little Colonel, Dimples,* and her own favorite, *Wee Willie Winkie,* in which she indulges her tomboy tendencies by wearing a uniform, carrying a wooden rifle, and running in front of a group of stampeding horses.

Over four consecutive years, from 1934 to 1938, Shirley Temple held her rank as the industry's top moneymaking star, displacing a multitude of grown-up luminaries, including Colbert, Gable, and the team of Fred Astaire and Ginger Rogers. International audiences swooned, too. In Russia, Shirley was the most beloved film star, along with Mickey Mouse. A reporter writing about her global appeal gushed that she took central Europe "like a full-sized meteor." Her movies broke box office records in the United States; in China the imperious Madame Chiang Kai-Shek couldn't get enough of Shirley's charm as a little orphan in *Curly Top*.

She had celestial appeal. Eight thousand children flocked to see the seven-year-old disembark from the S.S. *Mariposa* ocean liner for a holiday vacation in Honolulu. A department store Santa Claus asked for her autograph. And she attracted the world's luminaries to Hollywood, among them Irving Berlin and First Lady Eleanor Roosevelt. From her child's vantage point, she stared at a lot of belt buckles, handbags, and shoes, and sat on more famous laps than perhaps any other child in history. Some were bony and uncomfortable; J. Edgar Hoover's was "outstanding," she later recalled, because he didn't bounce or wiggle.

Kids wanted to *be* Shirley Temple—or at least be *like* her—and adults were happy to oblige. Parents named their babies after the star, boosting "Shirley" into one of the top five most popular names in the mid-1930s; Shirley Temple look-alike contests sprung up; Shirley Temple dolls filled store shelves; and Shirley Temple fan clubs amassed some four million members. Admirers sent several thousand letters to Shirley every week and even gifted live animals to the young zoophilist, including a Jersey calf from Oregon and two wallabies from Australia.

In 1935, the Academy of Motion Picture Arts and Sciences gave Shirley its first Juvenile Award. The presenter, journalist and humorist Irvin S. Cobb, asked her for a kiss before handing over the statuette—a request she remembered with little enthusiasm. One year later, *Who's Who in America* included Shirley as one of its newcomers, along with Nobel laureate Albert Einstein and high-altitude pilots Orvil A. Anderson and Albert William Stevens, who'd set a record of 72,395 feet in their helium-powered balloon.

Shirley's influence extended beyond popularity to economic solvency. Her triumphs on film saved a struggling Fox from near bankruptcy during the Depression. Thanks to their young star, whose salary rocketed from an initial $150 to $2,500 a week, Fox acquired the cachet to merge with 20th Century Pictures, creating the movie conglomerate 20th Century Fox. Moviegoers couldn't get enough.

Years later, child stars who acted alongside Shirley remembered her as a kid with talent, heart, and an easy rapport with the grown-ups around her. "She had some kind of magic *something* that other children simply didn't have," said her stand-in, Marilyn Granas.

She was "extraordinary," said Cesar Romero, who acted alongside her in several movies, including *Wee Willie Winkie* and *The Little Princess*. "She had a lot of guts, that little girl."

⌐⌐

What are the factors that transform a baby born on a spring day in California into the greatest child actress the world has ever known?

Much has been surmised about the role that Gertrude Temple played in her daughter's acting career. As early as March 1936, a *New York Times* story described the big questions on Hollywood tourists' minds: "They want to know whether Joe E. Brown's mouth is 'actually that big' and whether Miss Dietrich really wears trousers. W. C. Fields's nose is always a subject for speculation." But above all, the writer went on, they want to know about the most famous little girl in the world: "Probably the questions asked most often are: Is Shirley Temple spoiled? And is Mrs. Temple a typical movie mother?"

This question was not without merit. Judy Garland's mother, Ethel Gumm, gave her three actress daughters "pep pills" (better known as amphetamines) to keep them energetic for auditions and sleeping pills to make them drowsy at night, according to Garland's biographer Gerald Clarke. Soprano virtuoso Maria Callas rued the domineering role her mother, Evangelina Kalogeropoulos, played in her singing career. "I'll never forgive her for taking my childhood away," Callas told *Time* magazine in 1956. "During all the years I should have been playing and growing up, I was singing or making money."

No toddler decides to become a star; an adult must kick-start the process. Gertrude Temple made "a calculated decision" to enroll her daughter in dance lessons, Temple later recounted, as a way to harness her daughter's energy and cultivate her physical coordination. She took her to movie auditions and, once the film shoots began, stepped in as stylist and coach. Gertrude sewed her dresses, coiled Shirley's hair into 56 curls, and read entire scripts aloud so her daughter could memorize them while lying in bed with her eyes closed. "She reads and reads and reads," six-year-old Shirley told a *Los Angeles Times* reporter in 1934. "I talk and talk and talk."

Tall, willowy, and shy, Gertrude could come off as aloof at times. But Shirley's contemporaries also remember her as a dedicated and loving mother. "I adored Gertrude Temple. She was warm and generous to us and to my family," said child actor Dick Moore, who was famous for giving Shirley her first romantic kiss on-screen in the 1942 movie, *Miss Annie Rooney.* "And also I sensed in her somebody that wasn't going to be pushed around, and I envied that. I wished that my parents had been more that way."

Gertrude was certainly omnipresent. "Mrs. Temple would sit on a stool in back of the camera and she loved to see the different things that Shirley did," said Marcia Mae Jones, another child star of the era. "And she would always say, 'Sparkle, Shirley.' And Shirley sparkled all the time." Critics have charged that Gertrude may have been living her own dreams through her daughter and that she controlled Shirley's every move, denying her a normal childhood.

But Temple disavowed these concerns repeatedly, both in her memoir and in interviews. She described Gertrude as her partner, teacher, and best friend, neither forceful nor tyrannical. Her 1988

Spark

autobiography is a paean to her mother—she wrote it in large part to correct misperceptions—and although she admits to Gertrude being tough and intentional about her career, she describes her as loving and protective. When Larry King asked Temple if she ever felt pushed by her mother, she responded, "No, never. I think I pulled."

Gertrude didn't make Shirley enjoy the singing, dancing, and acting—or make her excel at it. As it happened, Shirley was equipped with abilities that helped her succeed, including two traits that have been documented in other prodigies: an exceptional working memory—the ability to not only remember information but also to hold onto it while doing something else—and remarkable attention to detail.

Shirley's memory allowed her to recall and recite her lines (as well as those of adult costars who slipped up) while acting and dancing on stage; her attention to detail played out in her ability to sense the heat of stage lights and master complicated dance steps. Above all, she seemed to have the "rage to master" that psychologist Ellen Winner discovered among art prodigies: They love what they do, they want to spend time doing it, and they do it well because they work hard.

This became exceptionally clear in a famous staircase dance that Shirley honed with Bill (Bojangles) Robinson in *The Little Colonel*. Robinson, who would become a lifelong friend, was good-natured, but demanding. "Every one of my taps had to ring crisp and clear in the best cadence. Otherwise I had to do it over," Shirley later recalled. The two rehearsed repeatedly until they stepped and scuffled and tapped in perfect harmony. "Practicing until each move became unthinking was a joy. Learning, an exhilaration," she wrote. "The smile on my face was not acting; I was ecstatic."

In an interview in 1937, during Shirley's domination at the box office, her closest rival Clark Gable joked that he couldn't compete with the three-foot-tall wonder with the lashes and smile. But he also bowed to her talent.

"She's not just a cute youngster. If it were just that, there'd be plenty of other cute youngsters to take her place," Gable said. "Well, can you figure anybody taking Shirley's place? People on the set have told me how she handles herself, how she never blows up . . . I'd probably find myself rehearsing in front of a mirror, trying to keep up with her."

On March 30, 1942—just three months after Japan attacked Pearl Harbor—*Life* magazine featured a portrait of Shirley on its cover with the headline "Shirley Temple Grows Up." Inside, readers were treated to portraits of the star, taken by Hollywood glamour photographer George Hurrell, and were informed that she was about to celebrate her 13th birthday.

The editors got it wrong. Years earlier, 20th Century Fox had shaved a year off Shirley's age and altered her birth certificate to make her appear even more precocious. In 1942, she was actually turning 14. By then, Shirley Temple had appeared in more than two dozen major motion pictures and advertisements ranging from Quaker Puffed Wheat and Sperry's Drifted Snow Flour to Dodge's 1936 sedan.

But by the time she reached adolescence, Shirley Temple's film stardom had started to dim. By 1939, the year she turned 11, she had dipped to number five in box office popularity, and Fox had become indecisive about how to cast their growing girl. Shirley's last film with Fox, *The Blue Bird*—an answer to competitor MGM's *Wizard of Oz* with Judy Garland—was a failure at the box office. *Time* summed it up succinctly: "*The Blue Bird* laid an egg."

Most importantly, Shirley's enthusiasm for the cutesy roles she played as a child was fading: Hollywood was beginning to feel stifling.

Spark

In late 1939, as she began aging out of Hollywood, Gertrude enrolled Shirley in the seventh-grade class at L.A.'s Westlake School for Girls. Until then, Frances Klampt (or "Klammy" as Shirley called her) had privately tutored her in Shirley's private bungalow on the Fox lot. An innovative teacher, Klammy capitalized on Shirley's interests. Rather than hand out geography worksheets, she and Shirley mapped Amelia Earhart's 1937 trek around the globe, from Timbuktu to New Guinea and, finally, Howland Island—a speck of land in the Pacific Ocean that Earhart was bound for when the plane she was flying with navigator Fred Noonan disappeared.

Surrounded by peers for the first time at Westlake, Shirley had to learn how to navigate the mystifying world of adolescent girls. After a rocky start as a celebrity newcomer, she settled in, making friends and soon checking off a host of accomplishments: class secretary, school newspaper reporter, and president of the Camp Fire Girls chapter, a scouting and leadership program.

While a student, Shirley did not give up acting entirely. In 1943, she signed with producer David Selznick and performed in the films *Since You Went Away* and *I'll Be Seeing You*. But real life became more enticing, and high school ignited a new passion: boys. Shirley kept a loose-leaf binder filled with photos of potential boyfriends and accomplished her goal of becoming the first senior to be engaged. On September 19, 1945, just a few months after graduating from high school, 17-year-old Shirley Temple married 24-year-old Sgt. John Agar, Jr., the brother of a classmate and a tall, blue-eyed Air Corps recruit.

Five hundred guests attended the wedding at Wilshire United Methodist Church, thousands of fans lined the streets, and wedding gifts poured in, including a 22-carat berry serving set with gold trim and crystal goblets. But the shimmer wore off quickly, as Temple described in her memoir. Just weeks into their marriage, Agar began drinking, told his new bride that he preferred long-legged model types, and slept through the delivery of their daughter, Linda Susan, born in January 1948. The arguments intensified. In October 1949, Shirley filed for divorce, retaining custody of their daughter, who goes by Susan.

Shirley refused to wallow in despair. On a trip to Hawaii with her parents in 1950, the year her divorce was finalized, she met Charlie Black at a cocktail party. A former naval officer and business associate at the Dole Pineapple Company, Black did not recognize the star and had never seen any of her films. His obliviousness charmed Shirley. "It was very refreshing to me—a handsome guy who wasn't interested in Hollywood or anything about it," she later said. "He was the love of my life."

Shirley's marriage to Charlie Black lasted 55 years, until his death in 2005, and produced two more children, Charles, Jr. (who also goes by Charlie), and Lori. Shirley's son tells me that his father was so loving toward his mother and so secure in himself that he never minded being called "Mr. Temple." Instead, he offered her a whole new life in Northern California, far from Hollywood and her earlier existence. "I'm quite certain her life would not have played out the way it did if she hadn't married this San Francisco boy," says Charlie.

The family of five ultimately settled in Woodside, California, some 350 miles north of Los Angeles. Here, on the San Francisco Peninsula, Shirley reveled in a relatively conventional life as wife, mother, and volunteer for a range of activities, including the National Multiple Sclerosis Society—a cause she would adopt after her brother George was diagnosed with the illness in 1952.

When she was 39 years old, Shirley Temple Black stepped outside of family life to embark on an entirely new challenge. After campaigning for Republican candidates in local and state elections, she decided to throw her hat in for a political run. On August 29, 1967, when her children were teenagers, she held a press conference announcing her candidacy for Congress in a special election to fill a seat in California's 11th district. Wearing a long coat and a jade necklace, Temple took on the Johnson administration and said she wanted to do her share to get the country "back on the road of progress."

The only female candidate in a crowded field, her opponents included a former skipper of a World War II PT boat, who quipped: "The campaign may shape up as PT-453 vs. the Good Ship Lollipop." By then, Shirley had been retired from Hollywood for almost two decades, but she would always be the little girl on stage. When reporters asked if her biggest challenge was being remembered as a child star, she said: "Little Shirley Temple is not running. If someone insists on pinning me with a label, make it read Shirley Temple Black, Republican Independent."

Shirley lost the election to a fellow Republican. But she was far from defeated. Over the next two decades, she would traverse the globe, serving as a highly regarded American diplomat.

"Long ago," she said in an interview, "I became more interested in the real world than in make-believe. I can hardly wait to see what happens next."

The third chapter of Shirley's life had just begun.

Prodigies are exceedingly susceptible to burnout. Once the adoration and fame wear off, self-esteem and confidence crumble. Many young actors struggle with addiction and mental health challenges as they fall from the heights of celebrity to the reality of everyday life. What is a person meant to do when eminence comes so early?

This is a challenge for any domain and any career, not just entertainment. Psychologist Dean Keith Simonton points to William James

Sidis, a math and linguistic prodigy born in 1898, as a window into this phenomenon. Sidis's father, Boris, a psychologist, became preoccupied with his son's intellectual prowess and accelerated his studies, pushing him to enroll at Harvard when he was just 11 years old. Billy was not yet a teenager when he gave a lecture on four-dimensional bodies before the university mathematics club.

But the boy's IQ was no defense against social ostracism. Sidis was bullied by students and hounded by the press, who couldn't get enough of the young mastermind. Eventually, after a brief stint as a teenage professor at Rice University, Sidis began studying at Harvard Law School and then mysteriously dropped out in his final semester. Over the next two and half decades, he took on a series of low-level accounting jobs and wrote several books under pseudonyms—including one on collecting streetcar transfers, his beloved hobby.

What Sidis wanted more than anything was to live in anonymity. He sued the *New Yorker* in 1938 for invasion of privacy and libel after it published a disparaging profile suggesting he had not lived up to his potential. In the spring of 1944, the *New Yorker* settled with him on the libel portion of his lawsuit. A few months later, Sidis died of a cerebral hemorrhage at the age of 46.

Prodigies who excel in intellectual pursuits, including math, music, and chess, "enter a world that far outstrips their social development," Simonton reports. A myopic focus means they are often physically and emotionally isolated from their peers, denying them the opportunity to learn much needed social skills. Their subjects are so abstract that they "may become lost in a private universe that prepares them little for practical living," Simonton writes.

To thrive, we must learn to be social beings, able to communicate with one another and the world. Toddlers figure this out as soon as they set foot in the playground sandbox. They need to share space and toys, and cooperate—as best they can—so they can navigate disagreements.

Children who develop these social and emotional skills early are more likely to be motivated to learn, do better in school, and grow into healthy and capable adults. Those who do not risk the fate of Billy Sidis.

Shirley Temple's capacity to flourish after child stardom was likely rooted in part by her ability to develop strong relationships. Although the studio kept her separated from others in her own bungalow, she spent much of her time interacting with other child actors, adults, and tutors. Asked how she was able to triumph at such a young age, she said: "I love people."

Her daily requirements as a toddler involved navigating social situations, both on and off the stage. Adult actors served as her first work colleagues, and she bonded, especially, with the crew, who became her playmates, mentors, and confidantes. "My cosmos was the studio and my universe the soundstage," she said. She would have paid Hollywood, she often noted, for the privilege of working.

Skeptics have long worried about the impact of employing children in such demanding jobs. But other child stars who thrived later say they appreciate the way Shirley viewed her past. "I completely understand loving it and feeling like you would have paid to do it," says the child actress Lisa Jakub, best known for her role as Lydia Hillard in the 1993 film *Mrs. Doubtfire,* starring Robin Williams. Jakub, who was cast in her first movie at the age of four, says acting provided her with a sense of community with like-minded people who work long hours together toward a common creative goal. "There's a little bit of magic there," she says.

Acting is also a natural extension of what children love to do, she says. "All kids pretend to have imaginary friends and all kinds of jobs and fantasy lives. They make themselves into kings and queens and animals," says Jakub, who retired from Hollywood when she was 22 and is now a writer and yoga teacher. "I think in a lot of ways, acting is more innate and natural for kids than it is for adults."

Shirley Temple

I wanted to learn more about how Shirley Temple pulled it off and I knew who would be able to help me understand. One bright fall afternoon, I make my way through the winding streets of hilly Woodside, California, where she lived for the last 53 years of her life.

Her children Charlie and Susan graciously invite me into her home. As we sit in the living room, they talk about their mother's love for them and for the two other most important people in her life—her mother and her husband, who adored her. "That was a real love story," Charlie said.

Their mother loved to cook, they tell me. She was funny, and warm, and loved decorating for the kids' birthday parties. Although immensely protective of her children, she also made a point of opening the family home to guests so that everyone would experience the joy she took in meeting people and learning about other places and other lives. When her children were little, Temple hosted a popular television series for kids, *Shirley Temple's Storybook*. They knew she was famous, but her priorities were always clear to them. "She was a normal mom," says Susan. "It was awfully nice to be her child," adds Charlie.

The living room embodies their mother's eclectic spirit—floral couches, vibrant red-orange carpeting, a 15th-century Spanish frieze along the walls. It all somehow gets along. In a corner is Black's long-standing artificial Christmas tree, lit up for my visit, and a symbol of their mother's favorite holiday. She loved the excitement; she loved giving presents; she loved the spirit of it all.

The private Shirley, their mother, was a deeply generous person with passionate views about the most important issues, Charlie tells me. Although a fiscally conservative Republican, she was socially moderate, he says, proudly pro-union and pro-choice. "She cared about the environment, she was against apartheid, she supported the Equal Rights Amendment," Charlie says. "She was very much an enlightened person."

As I listen to them talk, it becomes clear that of all the characteristics applied to their mother, one stands out above the rest: She was authentic. "What you saw was what you got," Charlie says, echoing the words of his father, who joked that his wife would have been "catastrophic" for the psychiatric profession. "Over 38 years I have participated in her life 24 hours a day through thick and thin, traumatic situations, exultant situations, and I feel she has only one personality," he told the *New York Times* in 1988. "You can wake her up in the middle of the night and she has the same personality everybody knows. What everybody has seen for 60 years is the bedrock."

Susan and Charlie believe their mother's strength came from the solid grounding she received in early childhood and in the family she created as an adult. She was "surrounded and loved," says Susan. Family fueled her from the beginning to the end.

When asked in 1988 who she'd choose if she could come back as anyone in another life, Shirley replied that she'd pick herself: "I cannot think of a more interesting life to ask for."

In the summer of 1968, when she was 40 years old, Shirley ventured to Prague as a volunteer to negotiate the Czech Republic's entry into the International Federation of Multiple Sclerosis Societies. On the morning of August 21, after completing her mission, she awoke to banging on her door at the Alcron Hotel. Russian tanks and troops were entering the city.

As a child star, Shirley had been exposed to danger. The kidnapping of baby Charles Lindbergh in 1932 spread fears about child safety, and she had experienced a number of perilous incidents, including extortion threats and a terrifying moment at a Christmas Eve radio benefit when a woman in the audience, who believed Shirley had claimed her daughter's soul, pointed a loaded gun at her before being tackled by security.

Prague exposed middle-aged Shirley to a different kind of danger. In a gripping journalistic account she wrote for *McCall's* magazine, she described a surge of people in the streets as tanks barreled into the city, bursts of machine gunfire, the body of a woman killed outside the entrance of the hotel she was staying in, and her own dangerous escape out of Prague to the border of Germany.

It was a pivotal moment in Shirley's adult life. Prague awakened her to international geopolitics, inspiring her to get involved in global affairs and crystallizing the life she wanted to create going forward. "She saw that as the beginning of it all," says her son, Charlie.

In 1969, President Nixon appointed Shirley as a delegate to the 24th United Nations General Assembly, kicking off her diplomatic career. It was an entirely new arena, but she found she had the know-how to tackle unfamiliar territory. "She said, OK, I've got to pull my socks up, I've got to educate myself quickly," Susan tells me. And she knew how to do it. "That goes back to going over scripts with her mom as a little kid. Those tools were carried through and just repurposed and used again," says Susan. She never dropped a useful tool, adds Charlie. "When she mastered something, there were arrows in her quiver."

Shirley's views were strong—she supported the Republican platform and strongly opposed communism—but she was also a gifted communicator who earned goodwill by working with her adversaries. In 1972, she served as a member of the American delegation to the UN Conference on the Human Environment in Stockholm, the first international gathering to address environmental challenges on a global scale. A series of diplomatic appointments followed. In 1974, President Gerald Ford tapped Shirley as ambassador to the Republic of Ghana and she later served as his chief of protocol at the State Department—

the first woman to hold that position. And in 1989, she began a three-year post as ambassador to then Czechoslovakia under President George H. W. Bush. Her husband, who spent most of his career in aquaculture, went with her.

Cold warriors questioned whether Shirley was up to the job, but she navigated the position with intelligence and skill, even during her first turbulent months when the country revolted for independence from communism. Her command of difficult situations was legendary. Norman Eisen, ambassador to the Czech Republic from 2011 to 2014, told *Time:* "She awed her costars with that quality on the set, and the same was true on the stage of world affairs."

Through it all, Shirley Temple Black balanced and intertwined the two ways she viewed the world: as both an optimist and a realist. These were qualities that allowed her to grapple with the many trials she endured over the course of her life. At age 22, she discovered that of the $3 million-plus she had earned, there was just $44,000 left in her trust fund. She didn't blame her father, chalking it up in part to bad advice from his business partner. "He was like an innocent led to slaughter," she told Larry King.

The truth was far more complex. As Temple noted herself in her memoir, her father played a significant role in mishandling her funds by not depositing them directly into her account. Some of her fortune went to expenses; her parents also helped out family members and loaned money to friends. Few paid them back. Still, Shirley accepted her fate. "My attitude has always been, get it over with, and get on with life," she wrote.

She took a pragmatic approach to doing so. "She used to say she was a person of action," her daughter Susan tells me. "She was not very often defeated." This became especially clear when she was diagnosed with breast cancer in 1972, at the age of 44. Writing in *McCall's,* she used her voice to educate others, urging women not to wait to be tested, "not to stay home and be afraid." She didn't sugarcoat the facts of how she was feeling (in pain and unattractive) and took a powerful stand on being in charge of choices about treatment. "The doctor can make the *in*cision, but I'll make the *de*cision," she wrote. Her honest account broke open the silence and stigma surrounding breast cancer and prompted a thunderous response. In the weeks that followed, readers sent Shirley hundreds of telegrams and more than 50,000 letters wishing her well, acknowledging their own diagnoses, and thanking her for speaking out.

"With or without a breast," she wrote, "I plan to keep doing precisely what I have been doing. Only better."

Over the course of her life, Shirley somehow knew how to mingle the child star with the grown-up she became. It was like a companionship, she once said. Yes, her childhood fame helped her later in life—nobody would let her separate the two even if she'd wanted to—but she had to succeed on her own. "The name, Shirley Temple, still opens doors for me," she liked to say. "But Shirley Temple Black has to perform or the doors will close."

And she did. Melissa Tonnessen, the teacher in New Jersey, discovered Shirley Temple in 1973, when Tonnessen was nine years old. One Saturday afternoon, she tuned into *The Little Princess* on television and was mesmerized. "I turned to my father, who was a movie buff, and said 'Who *is* this?'" she tells me. "It was a thunderstruck moment." In college, while her roommates tacked up posters of rock bands in their dorm room, Tonnessen hung one of *The Littlest Rebel*. Today, Tonnessen runs Shirley's Army, a Facebook group of more than 1,000 members who connect to share memories, show off Shirley souvenirs, and find joy in a turbulent world. Shirley Temple helped

Spark

Tonnessen through some unhappy moments in her childhood. She was an escape, a fantasy—a dream life Tonnessen wanted for herself. "She got me through hard times," says Tonnessen, noting that her infatuation with Shirley is as strong now as it was 50 years ago. "It was real then and it still is."

Shirley Temple Black spent her final days in her peaceful home in Woodside, a serene setting for the end of a remarkable life. On February 10, 2014, she died at the age of 85, surrounded by family. Obituaries hailed the arc of her life, from Hollywood stardom and prodigy to UN delegate and diplomat. She was, as the BBC reported, "that rare example of a Hollywood child star who, when the cameras stopped rolling, carved out a new career."

After we talk, Charlie and Susan invite me onto the sunlit patio out back behind Shirley's home, lush with orange trees and red bougainvillea dripping over the doorway. A wind chime sounds softly, a fountain of water gurgles, and a view east across the central San Francisco Bay peeks its way through a row of towering queen palm trees Shirley's husband planted for her years ago.

"She knew she was blessed," Charlie tells me. "We all knew we were blessed."

CHAPTER 3

Yo-Yo Ma

The Cellist (1955–)

"We shall now have the pleasure of hearing Yo-Yo Ma, accompanied by his sister Yeou-Cheng Ma, play the first movement of the Concertino No. 3 in A major by Jean-Baptiste Bréval," a dapper Leonard Bernstein announced at the National Guard Armory in Washington, D.C., on November 29, 1962. "Now here's a cultural image for you to ponder as you listen: a seven-year-old Chinese cellist playing old French music for his new American compatriots."

And what an image it was. Black-and-white film of that evening shows a diminutive Ma, who had lost a tooth just days before, approaching the stage with his 11-year-old sister and his cello, which reached somewhere around the tip of his nose. Yeou-Cheng wore a dress with shiny Mary Jane shoes, her hair pulled back neatly in a headband; Yo-Yo had on a suit and tie and a look of unflappable confidence.

The pair had come to perform at "An American Pageant of the Arts," a nationwide fundraiser for what would later become the John F. Kennedy Center for the Performing Arts on the banks of the Potomac River. In what the *New York Times* described as "one of the most ambitious closed-circuit television shows to be produced," the two-hour program was broadcast live from five locations: Washington, D.C.; New York; Chicago; Los Angeles; and a country club in Augusta, Georgia, where former president Eisenhower and his wife were vacationing.

Across the country, viewers paid $1 to watch from university auditoriums and movie theaters and as much as $100 for tickets to hotel banquets, where they watched and dined on roast lamb and filet mignon. The evening dazzled its collective audience of 350,000 with legendary artists, writers, and actors, including Bob Newhart, Colleen Dewhurst, and Jason Robards. Marian Anderson sang two spirituals; Van Cliburn played Liszt's Hungarian Rhapsody No. 12 in C-sharp minor; Robert Frost recited four poems; and Harry Belafonte serenaded the crowd with "Michael Row the Boat Ashore."

It was a remarkable evening of performances and a momentous debut for Yo-Yo Ma. Just months before, the young cellist had emigrated with his family from Paris to the United States and soon after had met the great Catalan cellist Pablo Casals: two of the most formative experiences of his life. It was Casals who brought Ma to the attention of the famed composer and conductor Leonard Bernstein. And now, here he was performing in Washington, D.C., in front of hundreds of dignitaries led by President Kennedy, dressed in a tux, and Mrs. Kennedy, in a white satin skirt and jacket with a pearl necklace.

As I watched the old footage online, it struck me: The theme of Kennedy's fundraiser would ignite the arc of Yo-Yo Ma's life. In his address to the nation that night, the president stressed the importance of poetry, art, and music during times of crisis and political strife. The arts are a social responsibility, he stressed, freeing the human spirit and reminding us that our desires and despairs are universal. "Today, as always, art knows no national boundaries," he said, and "genius can speak at any time and the entire world will hear it and listen."

In the six decades that followed, the young prodigy would transform himself from solo cellist to cultural emissary, traveling the world and crossing borders to unite people through music and dialogue. Early on, Ma's teacher Leonard Rose taught him to inhabit his instrument as if the strings were his voice and the body his lungs. In his playing, he has done that; in his life, the cello has become much more—an instrument for global fellowship, a way to engender compassion and build ties among people. "Knowledge in culture," he has said, "gives us perspective, and the capacity for empathy and humility."

Since his early childhood, Ma's life has evolved like the movements of his beloved Cello Suites by Bach—first the prelude, then the allemande, courante, sarabande, menuet, and gigue. Each decade has brought something new: He began playing cello at the age of four under the tutelage of his father, and gave his first public performance in Paris a year later. In his teens, he made his solo debut at Carnegie Hall and started college. In his 20s, he married, became a father, and started touring the world. Midlife brought musical collaborations— bluegrass, tango, jazz, the Silkroad Ensemble—and a determination to play with a purpose.

Today, in his 60s, Ma's mission is to harness the geography of hope in a fractured world. To do this, he performs, teaches, laughs, and listens. At a Bach concert at the National Cathedral in 2018, Ma addressed his audience. "I would like to dedicate the last two suites to all of you who try to do impossible things," he said, "who try to aim for something that's beyond what we think we can do. Because that's what Bach taught me on the cello."

Spark

⌣

Yo-Yo Ma's multicultural odyssey began the day he was born to Chinese parents in Paris. His father, Hiao-Tsiun Ma, grew up in Xinxiang, a village south of Shanghai, where he learned to play the violin and set out to pursue a career in music; his mother, Ya-Wen Lo, who later went by the name Marina, spent her childhood in Hong Kong before moving to China for her university education. The two met when Marina, a gifted singer, signed up for Hiao-Tsiun's music theory course at Central University, then located in Chongqing.

Born in 1923, Marina was 12 years younger than Hiao-Tsiun and smitten by her tall and scholarly professor who by then had traveled abroad and earned his doctorate in musicology at the Sorbonne. The couple had a shared love for music and decided to marry, settling in Paris after their wedding on July 17, 1949. The city provided new music, language, culture—and, it turned out, a refuge.

Just months after the couple was married, the Communist revolutionary Mao Zedong proclaimed himself chairman of the People's Republic of China, thrusting their homeland into decades of political upheaval. When Yo-Yo was born, on October 7, 1955, Mao was gearing up for what would become known as his Great Leap Forward, a movement that decimated the country's farmlands and led to massive famine and millions of deaths. Had the family not lived thousands of miles away in France, the trajectory of Yo-Yo Ma's life would have certainly been different.

Life in the Ma household in Paris was modest in means; Hiao-Tsiun made his living teaching music, and although Marina had trained in voice at the prestigious César Franck music school, tradition determined that she stay home and care for the family. The focus, from the start, was education. For Yo-Yo and his older sister, that meant homeschooled lessons taught by their father in a broad range of disciplines, from French history and mythology to Chinese calligraphy.

From the beginning, music permeated their lives like molecules dancing through the air—a life force that always, seamlessly, existed. They listened to music, they talked about music, they played music. Yeou-Cheng, who is four years older than her brother, showed exceptional skill on the violin. Yo-Yo wanted something different. When he was four years old, he spotted a double bass at the Paris Conservatory and announced to his parents: "I want to play that." They settled on a $^1/_{16}$-size cello. Although not much bigger than a full-size viola, Ma had to sit on three telephone books to play it.

Like Picasso, who plucked artistic imagery from his vast collection of memories, Ma is known for his ability not just to memorize music, but to also recall people's names and faces years after meeting in a crowd. Ma credits his father with training his memory to learn music, starting with the first piece he took on as a four-year-old cellist: the prelude to Bach's Cello Suite No. 1, a piece most young musicians don't attempt until well into their studies.

Hiao-Tsiun's trick was to break complex music into small, manageable parts. Every day, Yo-Yo was required to master two measures of a suite, building day by day until he had learned an entire movement. As he practiced, Ma began to recognize sounds and patterns, allowing him to understand the building blocks of composition. Rather than become overwhelmed by a profusion of notes over multiple pages, he calmly conquered musical phrases at a pace he could handle.

When he talks about this approach with interviewers, Ma often reaches for his cello, bowing one note slowly after the next to demonstrate how simple it is: Anyone can do it, he likes to say. But in addition to a private cello teacher, Ma had a disciplined instructor in his father, a sharp focus of his own, and a love for music that kept him engaged and eager for challenge. "By practicing only half an hour a day," he said in an interview, "I learned three Bach suites by heart by the time I was seven."

In 1962, the Ma family set out from Paris for New York. The journey was intended as a temporary visit with an urgent purpose: Hiao-Tsiun wanted to persuade his brother, who had recently moved from Shanghai to the United States to study electrical engineering, not to go back to China for fear of what might happen to him under communist rule. But during their stay, Hiao-Tsiun decided to accept a job teaching music in New York City, where he established the Children's Orchestra Society, which has since trained thousands of young students. Hiao-Tsiun's mission turned into a permanent relocation.

For Yo-Yo, the move from Paris to New York proved to be a seminal moment, opening his mind to a language, customs, sounds, and smells he had never before encountered. In Paris, he spoke Chinese at home and French in the city; now, he was living in New York City immersed in skyscrapers and the cultural melting pot of America.

One August day in 2019, I met Yo-Yo Ma at Tanglewood, home to the Tanglewood Music Festival in the Berkshire Mountains of western Massachusetts. I had come to interview Ma for a story for *National Geographic* magazine, and he greeted me warmly, as if I were a friend. At the start of our conversation, he talked about his childhood in Paris, his move from one country to another at an early age, and the profound impact of even the minutest differences he confronted. "The trees are manicured in France. In New York, there was Central Park, and you can walk on the grass. So think of what that does to a seven-year-old," he told me. "The cheese is wrapped in plastic. You have white bread as opposed to crunchy French bread and croissant."

Looking back, Ma regards this period as a time of confusion. Where did he fit in? Was he Chinese? French? American? But it was also an awakening, a realization that outside of his family lived a world of exciting unknowns. His parents, wedded to traditional Chinese ideals and expectations, focused on hard work and obedience. Now, as a young student in New York City in the early 1960s, he was encouraged to question, explore, and develop an independent mind. "In second grade, somebody asked, 'What do you think?' Nobody ever asked me what I thought," he told me. "This was incredible."

For Yo-Yo the prodigy cellist, New York offered new musical opportunities as well—beginning with a reference from Isaac Stern, who had heard Ma play in Paris at the advice of a famed luthier named Étienne Vatelot. "You should hear him," Vatelot told his friend. "He's unbelievable." The young musician also made a distinct impression on Stern, by then an esteemed violinist in his early 40s. "The cello he played looked larger than he was," Stern later recalled in his memoir. "I was astonished, truly astonished."

In New York, Stern arranged lessons with his chamber music partner, the legendary cellist Leonard Rose, with whom Yo-Yo began studying at the age of nine. "I was a pipsqueak of a kid, and over-whelmingly shy," Ma told the *New Yorker*. "I was afraid to speak to Mr. Rose above a whisper. I'd try to hide behind the cello." Under Rose's patient and nurturing tutelage at Julliard, Yo-Yo's musical skill and confidence developed. "By age twelve he could play some of the most difficult études and could remember everything he played by heart," Rose told journalist Helen Epstein in 1982. "He ate things up, and as he grew older he began to develop tremendous expressive qualities and musical intelligence. I love the boy; he has the ability to move me very much."

At home, Yo-Yo abided by his parents' strict upbringing. But he was restless and eager for independence. In his mid-teens, he spent a stretch of several weeks on his own for the first time at Meadowmount, a summer music camp in the Adirondacks. The experience was pro-found, allowing Ma the freedom to rebel against expectations. "I just went wild," Ma later said. "Never showed up at rehearsals, left my

Spark

cello out in the rain, beer bottles all over the room, midnight escapades to go swimming, and just about everything."

When he arrived back in the city to resume lessons with Rose, Yo-Yo wore a leather jacket and spouted some off-color language. Ever tolerant, Rose incorporated his student's newfound confidence into his teaching, granting Yo-Yo more leeway in his interpretation and expression of music—an unusual gift of independence early in his career that Ma recalls with gratitude.

On a Thursday evening in May 1971, 15-year-old Yo-Yo Ma made his solo debut at Carnegie Hall. His performance, "of a quality to make many an older man green with envy," the *New York Times* reported, included Beethoven's Sonata in A, Opus 69, and Chopin's Polonaise Brillante. Concerts around the country followed, earning him a flood of praiseworthy critiques: "electrifying," "staggering," "stunning." One critic noted that Ma possessed "both elegance and mercurial passion"; another, that "his command of the instrument is so thorough, and so effortless, that one half-imagines there was never a time that he didn't know how to play it."

As early as his teens, Yo-Yo was compared to the world's most elite cellists—János Starker, Mstislav Rostropovich, and Pablo Casals. This kind of adulation can be especially problematic for prodigies, whose lives are shaped by adults, largely in the absence of like-minded peers. Often, as they mature, they begin to question their own abilities. The countertenor prodigy Bejun Mehta experienced this in a stark way when his voice changed as a teenager and he felt a loss of identity that had been so tied into grown-up expectations. "I did not figure out as a prodigy," Mehta wrote in a poignant essay published in the 1998 book *Musical Prodigies,* "that the weight of adult opinion and expectation had helped me to lose sight of myself."

Ma somehow knew that he needed to explore a life beyond performance. He began university studies at Columbia while continuing his musical training with Rose. But it soon became clear that living at home with his parents would restrict his emotional and intellectual growth. He needed, in the words of Isaac Stern, "to become a person." In the fall of 1972, on the cusp of his 17th birthday, Ma enrolled at Harvard—a decision he marks as another formative moment in his life.

Ma is driven by curiosity; he hungers for information. You can see it in the way he stops and mingles with strangers at events, asking questions even as his managers tug on his shirtsleeves to move on. This spirit of inquiry found fertile ground at Harvard, where he immersed himself in a community of classmates from different hometowns, religions, and backgrounds. "Suddenly these people were living down the corridor from me and we were interacting," says Ma. They asked each other questions; they discussed their passions. These interactions, Ma told me, "set me off in life to realize how much there is in the world."

Grateful for his liberal arts education, Ma shares this lesson with anyone who will listen, cautioning talented young musicians and their parents not to stick to the narrow confines of music. What you put in your "emotional bank account" between the ages of 12 and 21, he says, is what you will withdraw from the rest of your life. At Harvard, Ma took courses in Dostoevsky, French civilization, fine arts, German literature, sociology, and discovered a passion for anthropology. The subject allowed him to study new cultures—including the !Kung tribe of the Kalahari, whom he would later visit—and to make sense of his own variegated cultural background.

In Cambridge, Ma gained another teacher and mentor, composer-conductor Leon Kirchner. He also began playing with two other Harvard musicians, violinist Lynn Chang and pianist Richard Kogan. The three became close friends, practicing and performing together as a trio. Kogan tells me he has strong memories of Ma's passion for learning, and he relishes the vivacity of their collaborations. "There

Spark

was an enormous sense of discovery; everything just felt so exciting," he says.

During his first year at Harvard, Ma spent almost every weekend performing. But he soon realized he was missing out on university life, and cut back to once a month. On campus, he played often with Chang and Kogan at Sanders Theatre. Ma made a point of sharing his talent, says Kogan: "He is incredibly generous." One night, upon their arrival for a concert, the trio found an overflow crowd who couldn't get seats. Rather than use the time to warm up backstage, Ma unpacked his instrument and started playing Bach's Cello Suites. "I was sitting there going over the score," says Kogan. "He sees people wanting to hear music and for him, that's always his first priority."

Now a concert pianist and psychiatrist at Weill Cornell Medical College in New York, Kogan studies the lives and minds of composers; in riveting performances, he weaves their stories with their music, unifying the musician and his art. Kogan tells me he is struck by the similarities between Mozart and Ma. Both had precocious talent and musician fathers who taught their sons with exacting discipline; both had older sisters who were also prodigiously talented musicians (Maria Anna Mozart, known as Nannerl, played the piano); and both avoided the perils of precocity that cause some young virtuosos to abandon their craft as they grow.

In Ma's case, says Kogan, the groundwork was laid early; he learned the fundamentals of technique so well as a child that he had the freedom to stretch his interpretations later on. "He has laser-like focus," says Kogan, but also the ability to break free when the music starts. "What's so amazing is that at the moment of the concert, there's a type of spontaneity that seizes hold of him. It's so contagious, you can't help but get caught up in the aura of it."

All of these qualities allowed Ma to develop both within and outside of his music, sparing him from the burnout that plagues young stars. "Music history is littered with stories of child prodigies who flamed out and never made substantial contributions as adults," says Kogan. "What's fascinating about both Yo-Yo and Mozart is that they made the transition from wunderkind to mature master."

⌒

Wolfgang Amadeus Mozart was born in Salzburg, Austria, on January 27, 1756, almost exactly 200 years before Yo-Yo Ma. In 1759, the three-year-old sat down at the clavier and learned to play. Within weeks of his fifth birthday—an age when most children are still mastering "Twinkle, Twinkle, Little Star"—the young boy composed his first pieces of music, an andante and an allegro for the keyboard. At seven, he embarked on what would become a three-year concert tour, where he showed off his remarkable musical accuracy and agility to European royalty.

Wolfgang's father and teacher, Leopold, hailed his son's talent— overzealously, some would argue—and devoted himself to cultivating the young boy's musical career. "Everyone is amazed," Leopold gushed in a letter after one of his young son's performances in Vienna, "and everyone whom I have heard says that his genius is incomprehensible." Wolfgang, his father later remarked, was a "miracle, which God has allowed to see the light in Salzburg."

Where does a musical prodigy's genius come from? Raw talent, hard work, ambition, a parent's resolute guidance—all certainly play a role. But is there something different about the way their minds operate? Is the brain of a prodigy unusual?

Research over the last several decades has found that gifted children in general score high on a variety of cognitive tests, including IQ, attention to detail, information processing, and working memory. Among musicians, the latter stands out as a critical feature, allowing performers to retrieve and process musical notes while also playing sonatas or études on their instruments.

Musical prodigies tend to show this ability early in life; often, they are able to reproduce music from memory before they learn to read it. And this skill develops the more it is practiced. Studies show that music training in childhood can alter the architecture of the brain, including changing the volume of the cerebellum, which is involved with working memory and has been found to play a role in learning music.

Understanding the origins of musical talent is a complex task that has led researchers to explore a wide range of populations. One of the most intriguing is a group of individuals known as savants, whose extraordinary abilities emerge despite—or perhaps because of—cognitive challenges.

When I call psychiatrist Darold Treffert, who has been studying savant syndrome for more than 50 years, he tells me the story of Leslie Lemke, who surprised his foster parents by playing Tchaikovsky's Piano Concerto No. 1 after hearing the piece on a movie sound track. Although blind and unable to read music, Lemke revealed a characteristic that is shared by other savants Treffert has studied: a prodigious memory. "He had this tremendous recollection of anything he heard one time," says Treffert, "and he would play it back, no matter the complexity or length."

Listening to Treffert talk about Lemke reinforces the astounding complexities of the brain. Neuroscientists who study it will tell you that they are on a formidable journey that has only just begun—think base camp on a climb to the peak of Mount Everest. Still, their research thus far has uncovered important features that are pertinent to Treffert's work and the study of human potential overall.

First, areas in the left and right hemispheres of the brain have particular roles; in simple terms, the left is more responsible for language, logic, and critical thinking, while the right tends to foster attention, emotion, and intuition. Second, despite these distinctions, we should in no way think of the left and right hemispheres as one versus the other. In a healthy brain, the two sides operate in concert through dynamic neural networks.

Based on this understanding, Treffert believes there are clues to how the unique abilities of savants emerge. Many have experienced damage to the left side of their brains, which explains their cognitive delays and their susceptibility to speech impairments. Treffert believes that savants like Lemke recruit from the right hemisphere, effectively rewiring their brains and freeing up creative capacities that might otherwise lie dormant—including the ability to decode melodies.

The minds of savants lay bare what Treffert calls "islands of genius"—a particular prowess that stands out in contrast to a disability. Treffert believes that savants are born with what he calls "genetic memory" for music, like monarch butterflies that somehow know they must fly from Canada to the jungles of Mexico every fall.

Savant syndrome prompts us to reconsider conventional assumptions about intelligence. About half of savants have autism, and the majority have IQs below 70 (Lemke's verbal IQ measured 58). This challenges the notion that a standardized test focused heavily on logic, math, and language skills is capable of capturing other kinds of brilliance. Savants who excel in art are able to draw with exceptional realism—as well as or even more effectively than art prodigies of the same age, according to the work of Ellen Winner and Jennifer Drake. And even art prodigies who are not savants do not have above-average IQs, despite their talent. "This is striking," they note.

Intelligence tests can both underestimate and overestimate potential. Despite common perceptions, genius and high IQ are not inextricably linked. In a landmark study launched in the 1920s, Stanford University psychologist Lewis Terman tracked more than 1,500 students with IQs largely greater than 140—a threshold

Spark

Terman labeled "near genius or genius"—over several decades. Many grew up to be highly accomplished professors, doctors, writers, politicians, and scientists. But others struggled to thrive, including several dozen who flunked out of college on their first attempt. Two students who were excluded from the study because their IQs did not make the cut, Luis Alvarez and William Shockley, grew up to win Nobel Prizes in Physics.

IQ tests depend on a person's responses to questions. But what about when we peer into their brains in real time? Scientists are finding intriguing results as they map the brains of musicians. In 2016, the musical artist Sting volunteered to go into an fMRI machine at the McGill University lab of neuroscientist Daniel Levitin while on tour in Montreal. The scans revealed that when Sting was asked to compose a fragment of a song in his mind, combining melody and rhythm, he activated a cluster of brain regions that were distinct from parts of the brain he relied on when writing a piece of prose or imagining that he was painting a canvas. Levitin was intrigued by a finding he didn't anticipate: Sting's brain responded similarly to unrelated music—for example, the Beatles' hit "Girl" and "Libertango," a tango piece by Argentinian Astor Piazzolla. Both are written in minor keys and have similar melodic motifs, suggesting that Sting is highly attuned to musical elements like tempo, key, and pitch.

Charles Limb, an auditory surgeon at the University of California, San Francisco (and a piano, sax, and bass player) is similarly fascinated by the neural effects of music. Limb has been studying the brains of musicians, including jazz artists and rappers, for more than a decade, so when classical pianist Gabriela Montero offered to submit to his fMRI machine a few years ago, he jumped at the chance. A child prodigy with perfect pitch, Montero began playing the piano before she could speak. She is unique, Limb tells me, because although she is a master of standard Western classical music repertoire, she truly shines when improvising.

Limb asked Montero, while lying in his scanner, to play music from memory and to improvise on a keyboard she held on her lap. In his study published in 2019, Limb reported that improvising led to widespread activity in her brain, involving areas linked to emotional engagement. He was most surprised to discover that Montero's visual cortex appeared to play an important functional role while she played improvised music; she somehow saw it in her mind's eye. "It's the closest I think I have come to looking at the brain of somebody who is a genius," Limb tells me.

Of course, these single findings must be interpreted cautiously, and there are no conclusions yet to draw. But as research builds, the neurobiology of musical prowess may one day reveal itself in more detail. Even now, scientists at the Université de Montréal and McGill University are attempting to recruit musical prodigies for a study. "We are just trying to understand how their brains make them learn so fast and make them exceptional," cognitive neuropsychologist and lead investigator Isabelle Peretz said in a radio interview.

⌒

Even as genes and biology carry potential for exceptional talent, a person's life experiences play a critical role in how it is expressed. Researchers who study genius have uncovered a fascinating environmental effect: Being an outsider can play a powerful role in later success. This raises intriguing questions about how Ma's experience as an immigrant might have influenced the trajectory of his life.

Some of history's greatest minds, it turns out, were immigrants who reinvented themselves in their new homelands. Studies conducted in the mid-20th century, after newcomers flooded American soil, found that second-generation immigrants made up 25 percent of distinguished scientists and 32 percent of eminent mathematicians. About 20 percent of MacArthur Foundation Fellows, who receive awards for creative promise, are foreign born. And since 1901, more than 100 of the 350 Nobel Prizes awarded to Americans have gone to immigrants and individuals born outside of the United States, according to an analysis by Boston University professor Adil Najam.

Immigrants can revitalize the collective imagination of an entire society, as Dean Keith Simonton found in a historical study of Japan between the years 580 and 1939. Simonton compared eras in which the island nation opened itself up—allowing citizens to leave and foreigners to come in—against eras in which they shut their borders. He found that historical periods with a flow of travelers and newcomers tended to boost creative achievement in a range of fields, from medicine and philosophy to poetry and painting. Isolation, by contrast, "eventually resulted in creative stagnation or decadence."

What is it about some immigrants that makes them excel? Researchers have found one key trait that stands out: openness to experience. In academic psychology, openness is one of five recognized personality traits, along with extraversion, agreeableness, conscientiousness, and neuroticism. People who are open to experience get excited about new ideas. They're intellectually inquisitive, curious, imaginative, and adaptive to change. And they appreciate art and beauty. Among the "big five" traits, openness is the most strongly associated with creative genius.

Ma talks often about how being an immigrant has shaped his outlook, his life, and the way he approaches music. When he arrived in the United States as a seven-year-old, he tried to make sense of his new world by holding on to his favorite French foods, places, and comic books. "I wanted life in Paris to continue, but I was also so excited to experience this new existence in New York," he said in a lecture he delivered at the Kennedy Center in 2013. Sorting out how to live these lives in parallel required a crucial skill: imagination. He needed to figure out how to rise above geographic boundaries—a childhood pursuit that continues to be a lifelong quest.

Over the decades, Ma's life evolved into a road map of new discoveries, a quest driven by his yearning for knowledge. His early years taught him that life could change quickly and that he would figure out not only how to adapt, but also to thrive. His courses and interactions at Harvard made clear that the universe was rich with ideas, cultures, languages, music, literature, and history waiting to be explored. "That dovetailed well with the burgeoning career of a young musician trying to make his way in the world," Ma told me.

In his 20s, Ma began traveling internationally on tour, performing in Lapland one day and Frankfurt the next. He also endured risky surgery for scoliosis, a congenital curvature of the spine; his months-long recovery in a body cast kept him from practicing, but he recovered fully and returned to the stage with remarkable stamina. By 1983, when the 27-year-old made his debut with the Boston Symphony Orchestra—the place that would later become his home—Ma was playing up to 120 concerts a year. Although he was by then touted as the world's greatest living cellist, Ma has little appetite for titles—"prodigy," "genius," "superstar"—preferring instead to focus on what he can learn from everyone else.

In his 30s, Ma expanded his musical reach, launching unconventional collaborations with musicians in other genres. Jazz vocalist Bobby McFerrin challenged Ma to improvise—an experience the cellist described as "terrifying as well as exhilarating." It was also enormously successful. Fans relished the combination of disparate voices and bought over half a million copies of the duo's 1992 album, *Hush*.

One year later, Ma set out for the Kalahari Desert to study the musical traditions of the Bushmen he had learned about in college.

"This is probably the greatest stretch I've ever made," he said in a 1993 documentary he made about the experience. Even with vast differences in language and culture, Ma and the villagers communicated through music. And the cellist took home the lessons he learned: Music need not be formally performed in a concert hall—it can evolve at any place and at any time—and when it is passed down through oral tradition, it changes shape as it's played. "I'm going to try and play different kinds of music as a result of coming here," he said.

When he was close to 40, Ma decided to study bluegrass in Nashville with fiddler Mark O'Connor and bassist Edgar Meyer. In 1996, the trio produced what *Variety* called the "delicious" CD *Appalachia Waltz,* and a performance of their collaboration that year won raves. "Ma's gorgeous lyrical lines were his particular forte," the reviewer noted, "but he also made laudable efforts to transcend the printed music and get down like his friends." Then there was tango in Argentina ("The music hit me like a fever that wouldn't let go," Ma said) and the launch of his famed Silkroad Ensemble, in which he joined musicians from more than a dozen countries, spanning the globe from Syria and Japan to Spain and the United States.

But even while piling up the accolades and raking in the Grammys (he has won 18 so far), Ma questioned his existence as a musician. He was born into music and he was good at it. But he never chose to become a cellist—it was simply the person he became. What if he'd decided to be a teacher? A social worker? An anthropologist? At one point, he says, his longtime friend and collaborator, the pianist Emanuel Ax, said to him, "Yo-Yo, you've just got to stop thinking that your profession is an interruption of your life."

At midlife, when he was 49 years old, Ma had a revelation: His true passion is people. Why do they think the way they do? Why do they do the things they do? What drives them and interests them and moves them? Suddenly, his world made sense. All that confusion from childhood, the mix of cultures and search for identity, the second-guessing about his life as a musician—it all led to a truth that gave him meaning and sustenance: Music is his offering, a barter for human interaction. In exchange, he can get to know people, explore human nature, and understand the world.

On the night that Yo-Yo Ma, the newly arrived seven-year-old immigrant, played in Washington, D.C., many important people were present to hear him. But it wasn't President Kennedy or Jackie Kennedy or Eisenhower or Leonard Bernstein who made an impression on Yo-Yo. It was the actor, Danny Kaye.

In his 2013 lecture at the Kennedy Center, Ma made a point of showing a photo of the actor kneeling down in front of Yo-Yo and his sister. "Look at where his eyes are. He came down to my level in order to be an equal to the toothless wonder," he said to the audience. "He extended himself, met me at the crucial edge that divides adult from child, and he won my heart. Since then, I have subliminally internalized that gesture and that attitude. And today, I try to be mindful of this in everything I do—to meet people at eye level, at the edge that divides one person from another."

Ma starts with the divide between children and adults. He often says he is most proud of appearing on *Sesame Street* and *Mister Rogers' Neighborhood,* because it allowed him to become a guest in a child's world—and, perhaps, to have an impact. He makes a point of pushing past the crush of nervousness young musicians often feel. Graham Cullen, a cellist who studied with Ma in a master class at Tanglewood, says most teachers start their sessions by listening to students play and then providing instruction. Cullen was amazed that Ma took the time to get to know his students before anyone picked up a bow. "He asked everyone's name, where we're from, what we do," Cullen says. "He seemed like he really wanted to know, which is not usually the case."

Ma is often self-deprecating, kicking over a pedestal he'd rather not stand on. He likes to tell this joke:

A cellist walks on a beach and picks up a bottle. A genie pops out and says, "I'll give you two wishes."

The cellist says: "Wow, I'd like to have world peace."

The genie thinks for a second and says, "That's too hard! What's your second wish?"

The cellist says, "Well, I'm turning 60 and I want to play in tune."

The genie thinks for a second and says, "What was your first wish again?"

Once, on a walk through Harvard Yard with a *New York Times* reporter some years after graduation, Ma made a point of noting all the places where he'd faltered: "'That's where I almost failed German,' he said, pointing to one building. 'That's where I used to sleep in the stacks,' he added of another." Kogan says Ma has done this for as long as he can remember so that he can put himself on equal footing with everyone around him: "He's constantly looking to bridge whatever gulf or gap there is."

When President Barack Obama introduced Ma before awarding him a Presidential Medal of Freedom in 2011, he said the cellist was unique among stars: He likes to hug people; he nicknamed one of his cellos "Petunia"; he even won *People* magazine's Sexiest Classical Musician award in 2001. "But maybe the most amazing thing about Yo-Yo Ma is that everybody likes him," Obama said, before joking, "you've got to give me some tips."

Ma credits his hero, Pablo Casals, with shaping his approach to music and life. When Ma first played for him as a child, the virtuoso said, "Hmm, very good," Ma recalled. And then he added: "Make sure you have time to play baseball." Casals's playing awed Ma; "his musical phrasing had the strength and beauty of marble sculptures," he later said. But how Casals viewed himself affected Ma even more: "He said, 'I am a human being first, a musician second, and a cellist third.'"

Ma was 18 when Casals died at the age of 96 in 1973. But the elder cellist left his protégé a musical legacy: Bach's Cello Suites. Casals discovered the music, a bundle of discolored scores, inside a musty shop near the Barcelona harbor when he was 13 years old. Written around 1720, the solo music for cello had been relegated to academic studies until Casals recognized their value. "I had never heard of the existence of the suites; nobody—not even my teachers—had ever mentioned them to me," Casals wrote in his memoir. "I hurried home, clutching the suites as if they were the crown jewels." For the next 12 years, he studied them every day and became the first cellist to play one of them in its entirety in public.

Among cellists, Bach's Cello Suites have become the canon's pinnacle. For Ma, they are like a baby's first lullaby, imprinted on his soul. He has recorded the entire set of suites three times (in his 20s, 40s, and early 60s) and has played movements at weddings and funerals (for Senator Edward Kennedy, Katharine Graham, Steve Jobs). They have made their way to curious villagers in the Kalahari Desert and to grieving family members at the dedication of the 9/11 memorial in New York City. In 1991, Ma played the sarabande from the Fifth Suite at the bedside of his father before his death at the age of 80.

Over time, the suites—which become more challenging between first and last—have become something of a metaphor for Ma, because they require a leap of technical skill and imagination. In the Fifth Suite, Bach tuned down one of the strings, taking it from an A to a G; in the sixth, he composed music for a five-string instrument, requiring cellists to create a polyphonic sound on a single cello. The latter requires wizardry on the part of the musician, as well as participation from the audience. Listeners must sustain notes in their ears that the cellist cannot complete, Ma says—a partnership that enriches both parties. By working together, Ma has said, "we have the chance, for a brief moment, to be in touch with the sublime."

In 2018, the year he turned 63, Ma embarked on his most ambitious endeavor to date: the Bach Project, a two-year tour playing the Cello Suites in 36 sites on six continents. At each location—from Mumbai and Mexico City to Vienna and Christchurch—Ma played the suites and then dedicated additional time to a "Day of Action," where he engaged with local neighborhood leaders, citizens, students, and activists.

When I interviewed Ma for my magazine article, he talked about the power of music to bring people together in this kind of venue. It's an experience of sharing, like the pot in the classic children's folktale "Stone Soup," he told me. "I play the cello. This is the best of what I can bring to you. It's the best I can offer," he said. "What would you like to put in the pot? How would you like to start a conversation? What are the things you're thinking about? What are your needs— what is it that you're struggling with?"

At the National Cathedral in Washington, D.C., one of his first stops on the tour, Ma smiled and blew kisses to his audience before sitting down to play. Sometimes he threw his head back and closed his eyes; other times, he seemed to lock eyes with a single listener and smile. Rarely did he look at his left hand, his slim fingers pressing down on the strings, or at his bow, which moved between the bridge and the fingerboard. Halfway through, he raised his cello over his head like a barbell, acknowledged his "sore tush," and encouraged attendees to take a break. Then he did jumping jacks.

Ma plays the suites straight through without intermission, a feat that lasts about 2½ hours. It is hard to imagine the stamina required— the strength of mind, arms, fingers, and torso. And yet Ma bounds with energy. At his Day of Action in Pittsfield, Massachusetts, near Tanglewood, he arrived at a community event with arms outstretched in greeting. He visited with community leaders and townspeople who had come to build tables in a communal team-building event, planted a tulip tree, and insisted on giving spoonfuls of ice cream from his cup to his Tanglewood students, including violist Alaina Rea. "He's one of a kind, because he's obviously a great technical player, but he also has such an energy around him," Rea tells me. "His openness and willingness to share—that's his whole persona."

As I watched Ma interact, a word lit up in my mind like a movie marquee: "exuberant." It is a term, I later discovered, that has profound meaning to clinical psychologist Kay Redfield Jamison, who has written an entire book on the topic. Exuberance, as Jamison describes it, is an amalgam of optimism, energy, enthusiasm, and extraversion. Exuberant people engage energetically with the world and "hold their ideas with passion and delight," she writes. "Their love of life and of adventure is palpable."

Exuberant personalities, like Louis Armstrong, P. T. Barnum, and the beloved (even if fictionalized) Mary Poppins are like emotional bandleaders—skilled at expressing their emotions and transmitting those feelings to others. And exuberance is contagious, Jamison affirms, drawing people together and creating a community of hopefulness and promise.

The more I read about it, the more it seemed to fit. Exuberant.

⌒⌒

Over the course of his 60-year career, Ma has played for eight presidents and sold 10 million albums worldwide. Among countless other honors, he won the Avery Fisher Prize in his 20s, the Glenn Gould Prize in his 40s, and an Honorary Doctorate of Music from the University of Oxford in his 60s. He is also a husband, father, and grandfather. He and his wife, Jill Hornor, an arts consultant, met as teenagers at the Marlboro Music Festival in Vermont, and have two grown children: a son and a daughter who recently started a family of her own.

In the spring of 2020, when the global pandemic disrupted the last phase of his Bach Project tour, Ma took to the internet. He posted videos of himself playing the cello on Facebook and YouTube, and encouraged other musicians and nonmusicians alike to share their own #SongsOfComfort. On Memorial Day weekend, while most Americans remained isolated in their homes, he turned again to Bach, playing the Cello Suites live from inside the television stations of WGBH in Boston, where his performance was recorded by robotic cameras and streamed to the world.

In his role as musical ambassador, Ma has taken on the mantle of his hero, Casals. The tiny seven-year-old prodigy who bowed before the president in Washington, D.C., is now a grown man imparting his philosophy for a moral life to generations that follow. In 2019, when addressing the graduating class of Dartmouth College, Ma told students never to abuse their power. "Remember always that you are a human being first," he said. "Practice your humanity daily. Practice that truth. Let it power your decisions, let it inspire your thoughts, and let it shape your ideals. Then you will soar. You will fly. And you will help others soar and fly."

CHAPTER 4

Bill Gates

The Software Pioneer (1955–)

In 1967, when he was in the sixth grade, a very bright Bill Gates struggled to stay focused and disciplined. Thinking he might benefit from a different kind of learning environment, his parents, Bill Sr. and Mary Gates, suggested he move from the public elementary school he was attending in Seattle to Lakeside, a private boys' school near the shores of Lake Washington.

Bill, who would be turning 12 that fall, wasn't so sure. Lakeside required students to wear jackets and ties, attend chapel every morning, and call their teachers "Master." He thought about intentionally botching the entry exam so the school would reject him. But he couldn't help himself. He got in.

Decades later, when Gates returned to Lakeside to donate $40 million to its financial aid program in 2005, he credited the school with being "one of the best things that ever happened to me." At

Lakeside, after all, the inquisitive introvert collided with the dawning of the digital revolution, launching his breakthrough career in computer programming. "If there had been no Lakeside, there would have been no Microsoft," Gates said. "And I'm here to say thank you."

Gates's introduction to computers happened one fall day in 1968, when he and his fellow classmates walked into a small room in the basement of Lakeside's McAllister Hall and saw a Teletype terminal. The machine, which had been funded by a Lakeside Mothers' Club rummage sale, consisted of a console, a roll of paper, and a clacking keyboard so loud the walls and ceiling of the room had to be lined in corkboard. In his 1995 book, *The Road Ahead*, Gates described the terminal as "huge and cumbersome and slow and absolutely compelling."

By then an eighth grader, Bill was immediately drawn to the potential of the machine and the power of its user. All he had to do was enter the right combination of numbers and symbols and the terminal would take his orders and spit out a result. At the age of 13, he wrote his first programming code for a version of tic-tac-toe. Nothing could beat the thrill of typing his instructions into the keyboard, hitting "RUN," and waiting for the answer to come chugging back.

The idea that Bill could design his own software—the programming that tells computer hardware what to do—was intoxicating for a young teenager who loved numbers and excelled at math. "We were too young to drive or to do any of the other fun-seeming activities, but we could give this big machine orders and it would always obey," he writes. "That was the beginning of my fascination with software."

In the great lottery of human existence, Gates won big. He was born to parents who provided him with a privileged education and lessons in how to build a worthy and generous life. He had an intellectual aptitude that flourished when challenged. He came of age just as the personal computer morphed from futuristic pipe dream to technological reality. And, above all, he found partners at pivotal junctions, who brainstormed, debated, and joined forces with him.

Just weeks after his 20th birthday, Gates and Paul Allen, a fellow Lakeside student and computer wunderkind, officially registered Microsoft, the software company that made Gates a billionaire with its ubiquitous Windows operating system. In 1987, the year he turned 32, Gates met Melinda French, his future wife; together, they would discover a passion for philanthropy and form the Bill & Melinda Gates Foundation, which Gates Sr. helped guide and direct until his death in 2020. And in 1991, Gates made the acquaintance of investor Warren Buffett, who quickly became a lifelong adviser and friend, despite their 25-year age difference. These pairings propelled the arc of Gates's illustrious achievements.

Now in his 60s, Gates the philanthropist finds satisfaction in work that shares common ground with his childhood passion and his digital career. "It's about using innovation to solve problems and help people," he tells me. "And in both cases, I get to bring together smart people and collaborate with them."

"It is unusual to have so much luck in one life," he once said, "but it's been a major factor in what I have been able to do."

Given his passion for numbers, it seems fitting that Bill Gates was given a numerical nickname as a kid: Trey. Derived from *tres,* the Latin word for "three," "trey" is used to distinguish cards or dice with the rank of three, and to differentiate the third male in a line of family members with the same name. In Gates's case, his father went by

Spark

William Henry Gates, Jr., making him William Henry Gates III. To avoid confusion, his maternal grandmother, Adelle (Gami to her grandchildren) and Adelle's mother, Lala—both of them avid card players—suggested Bill III be known as Trey.

Born on October 28, 1955, in Seattle, Washington, young Trey landed in a family with solid financial security and social prominence. His father came from entrepreneurial roots seeded by his grandparents, who had moved to the Seattle area from Pennsylvania in the booming 1880s, when job opportunities beckoned intrepid Americans out west. Gates's mother, Mary Maxwell, grew up in a wealthy banking family renowned for its dedication to public service—a tradition she exemplified in her civic leadership and instilled in her children.

Bill Gates, Jr., met Mary Maxwell in the 1940s at the University of Washington, which he attended under the GI Bill after earning first lieutenant rank in the Army; he later earned a law degree. At six feet seven inches, he towered over his five-foot-six wife-to-be, but she convinced him she was tall enough for a first date by standing on her tiptoes. Gates's father was somewhat reserved; his mother, a popular "bundle of energy," a classmate later recalled. In his memoir, *Showing Up for Life,* the older Gates wrote with affection about Mary's intelligence, affability, and enthusiasm. After marrying in 1951, the couple settled down in Seattle, where Bill began rising up the legal ranks and Mary stepped away from teaching to raise their three children.

Trey was a smiley, upbeat kid. As a baby, he figured out how to rock himself back and forth in his cradle and swayed on a rocking horse when he was big enough to balance. Motion appealed to him, wherever he could find it. A home video, aired in the 2019 Netflix feature, *Inside Bill's Brain,* shows him beaming gleefully as he teeters on a balance board in front of the fireplace. He's dressed in a long-sleeve shirt, checkered overalls, and white shoes, and his cheeks are red, warmed by the flames burning behind him.

Gates's lifelong passion for reading began in his partially above-ground basement bedroom in the family's home in Seattle. When he was eight, he set out to conquer every volume of the 1960 *World Book Encyclopedia* for fun (he got as far as the *P*'s.) He also devoured sci-fi novels, including Robert Heinlein's *The Moon Is a Harsh Mistress,* one of his favorites. "I was lucky in that my parents would buy me any book I wanted," he told me.

Reading was how young Bill loved to learn. Family members remember him sitting in his cluttered room surrounded by tomes. He read at the dinner table, he read late at night, he read voraciously for the school's summer book competition—an activity "he always wanted to win and often did," his father noted.

Gates was not allowed to withdraw completely into a solitary existence, however. His mother and father—who would later adopt the name Bill Gates, Sr., to distinguish himself from his son—enrolled him in the Cub Scouts and made sure he engaged in the social rituals built into their family life. They watched the *Ed Sullivan Show,* played cards after dinner (the winner was exempted from helping with the dishes), shared Sunday roast beef dinners at Gami's house, and wore matching pajamas on Christmas morning.

Every summer, the family set out for its most treasured tradition: a getaway with eight other families at Cheerio, a rustic resort on the Hood Canal along Washington's Olympic Peninsula. Over the course of two weeks, the kids swam, played tennis, went water-skiing, and competed in "Olympic Games" that ranged from three-legged gunnysack races to egg-in-spoon sprints.

Gates reveled in the contests. Cheerio rituals also included pancakes on Sunday mornings and marching in an annual parade led by "mayor" Gates Sr. while belting out a camp ditty sung to the tune of the theme song from *Bridge on the River Kwai*. At Cheerio, Gates became a tennis enthusiast and watched Neil Armstrong step onto the surface of the moon in July 1969.

While Cheerio brought out the fun in him, Gates's tendency to cloister himself in his bedroom at home concerned his parents, as did the defiance that began to surface as he approached his teenage years. Gates's father remembered one day when his wife called out, "Trey, what are you doing down there?" to which his son responded, "I'm thinking, mother. Don't *you* ever think?" In one incident, deeply etched in family lore, Bill Sr. threw a glass of water in his son's face after he shouted at his mother at the dinner table.

When he was 12, Gates's parents sent him to a psychologist who convinced him that pushing back disrespectfully was a battle not worth fighting. Although his rebelliousness persisted—he would muster it later in college when he skipped classes, and use it tactically when launching Microsoft—he realized the therapist was probably right. "I definitely gave my mom a hard time not following along with what she wanted and really trying to detach myself," Gates told Netflix filmmaker Davis Guggenheim. "Just disobedience, disrespect. I mean, it's kind of embarrassing to think about it now."

Lakeside provided the intellectual stimulation he required. Instructors, like his English teacher Ann Stephens, refused to let him coast. One day, after noticing that he needed extra assignments to fill his time, Stephens handed Gates her 10 favorite books and her college thesis, and told him to read them. "Lakeside had the kind of teachers who would come to me, even when I was getting straight A's, and say: When are you going to start applying yourself?" said Gates during his 2005 visit to the school.

Most important for Gates, Lakeside had the means and the motivation to take a chance on technology, as well as the willingness to let students investigate its potential. Until the late 1960s, computers existed largely within government offices, private companies, and universities that could afford them. The Lakeside terminal offered students a chance to take part in a new venture known as time-sharing; this allowed their school machine to connect over a dial-up network to a General Electric mainframe in downtown Seattle. Fred Wright, a Lakeside math teacher, provided students with a manual that described how to program in BASIC (Beginner's All-purpose Symbolic Instruction Code) and unleashed his young computer scientists to explore the new technology on their own, like adventurers making their way through thickets in a jungle.

I wondered what it felt like for Gates to happen upon his love of programming as a young teenager, and asked him about it. "At the time, middle school was early in life to discover a passion for coding, because computers were so new and their capabilities were still pretty limited," he explained. "But for me, it was a lot of fun. And it's also how my closest friends and I spent our time together."

Among the flock of young techies who congregated in McAllister Hall was a strapping sophomore with sideburns and glasses named Paul Allen. He would become Gates's most enduring companion. A snapshot of the pair shows Allen, almost three years older, in a corduroy blazer sitting at the Teletype keyboard with Gates standing next to him in a buttoned-up plaid shirt.

In Allen's memoir, *Idea Man*, the cooler and more sociable upperclassman described his first impression of Gates: "One day early that fall, I saw a gangly, freckle-faced eighth-grader edging his way into the crowd around the Teletype, all arms and legs and nervous energy," he wrote. Bill was decked in a pullover sweater, khakis, and oversize shoes and clutched a marker in his mouth when focusing intently. "You could tell three things about Bill Gates pretty quickly," Allen observed. "He was really smart. He was really competitive; he wanted to *show* you how smart he was. And he was really, really persistent."

Although Gates and Allen differed in fundamental ways—their age, size, personalities—they shared an unwavering zeal to master the new technology that seduced them. Like an older brother, Allen would provoke Gates and push him to prove his proficiency. "Once in a while, when we got stuck on a problem, Paul Allen would turn to me and say: 'If you think you're so smart, you figure this out,'" Gates recalled. "And I would take those manuals home and read them page by page, over and over."

As their skills developed, the pair's ambition grew. Soon, Allen, Gates, and two other devotees, Ric Weiland and Gates's closest friend Kent Evans, formed the Lakeside Programmers Group and sought out real-life jobs outside of school. A local company hired them to look for bugs in their computer's new operating system. Although the boys didn't get paid, they got unlimited free time to tinker off-hours, along with an irresistible directive: to make the system crash. They worked for hours at a time after school and on weekends, reveling in their access to a more sophisticated machine—an upgrade that Allen later compared to going from a Corolla to a Ferrari. Gates was so hooked he would sneak out of the house at night to keep at it. "Mary and I had no idea our son was out late on school nights living a second life as a hacker," his father later wrote.

Gates was unlike his older sister, Kristi, directed and self-disciplined, and his younger sister, Libby, a talented athlete who inherited her mother's social ease. Growing up, Libby remembered thinking, "he's kind of weird, and his friends are weird. They're kind of nerdy and they think differently from me."

At one point, Bill Sr. and Mary became so worried about how much time their son was spending on programming that they ordered him to take a break from Lakeside's computer lab, according to biographers James Wallace and Jim Erickson. For about nine months, he stayed unplugged and spent his time consuming business and science books, biographies of Napoleon and FDR, and novels, including *A Separate Peace* and *Catcher in the Rye.*

But his interest in computers never dissipated. Already, he was thinking about how he and Kent Evans would go into business together. Kent had introduced Gates to *Fortune* magazine, which he carried around in his big lawyerly briefcase, and challenged his classmate to think big. They talked endlessly, brainstorming about how one day they would bring computers to the public.

In May 1972, Gates received devastating news: During the final ascent of a mountain-climbing course over Memorial Day weekend, Kent had slipped and fallen to his death on Mount Shuksan in North Cascades National Park. He was 17 years old. Gates was scheduled to speak at Kent's funeral but was too overwhelmed. His loss was profound. Even decades later, Gates still remembers Kent's phone number.

Back at Lakeside, Bill needed help. He and Evans had signed on to tackle an arduous project that summer. The school needed its class schedules computerized for the next academic session—a gargantuan task, not only because multiple classes and pupils had to be navigated, but also because Lakeside had just merged with a local girls' institution and doubled in size to 400 students.

Gates reached out to Allen, who was by then an undergraduate at Washington State University, and asked if he would lend a hand. Until then, the two had spent much of their computer time with others; the new venture meant they could work closely one on one, allowing them to develop a kind of "high-bandwidth communication," as Allen later described it, which laid the foundation for their Microsoft partnership. As they toiled away that summer, sometimes sleeping on cots on campus, Allen was impressed by Gates's ability to navigate the many moving parts, from class size to teacher breaks. And by his ingenuity: Gates would later note that he made sure to schedule "a disproportionate number of interesting girls in all my classes."

Spark

In the fall of 1973, as Watergate churned in Washington and Billie Jean King beat Bobby Riggs in "The Battle of the Sexes" in Houston, Gates entered Harvard as a freshman. He loved the intellectual atmosphere and the exhilarating challenge. But he also played a lot of poker, stayed up late, and became, as he later said, "the leader of the anti-social group." At one point, he decided to attend classes he had not signed up for and skip those he had. "My friends from Brain Studies thought it was very strange that I sat on the wrong side of the table and took the Combinatorics exam even when I was the most vocal student in the Brain class," he later joked.

Gates and Allen, meanwhile, never stopped fantasizing about their next software scheme. Before Gates left Seattle for Harvard, they had designed a program that would allow engineers to study traffic flow—their first official business partnership–which they named Traf-O-Data. They believed they were on the precipice of change. In the summer of 1974, Gates persuaded Allen to take a break from Washington State—he'd already taken two and had yet to graduate—and move to Cambridge. Allen got a job at Honeywell while he and Gates continued hunting for opportunity.

Destiny intervened one cold, snowy day that December when Allen went for a browse at the Out of Town News kiosk in the middle of Harvard Square. The headline on the cover of *Popular Electronics* magazine's January issue leapt out at him: "Project Breakthrough!" The inside story featured a home mini-computer known as the Altair 8800, made by an Albuquerque-based company called MITS (Micro Instrumentation and Telemetry Systems), which could be built for under $400.

Here, finally, was their moment. Allen shelled out 75 cents, grabbed the magazine, staggered through the slush to Gates's dorm room at Currier House, and handed him the story. The Altair would need software to function—a digital language the computer would understand—and Allen and Gates knew they were capable of creating it. "If we'd been older or known better, Bill and I might have been put off by the task in front of us," Allen later reflected. "But we were young and green enough to believe that we just might pull it off."

Over the next eight weeks, Gates and Allen, then 19 and 22, set up camp in Harvard's Aiken computation lab on Oxford Street, breaking for pizza and an occasional pupu platter from Aku Aku, a nearby restaurant, to keep them fueled. They hired a Harvard freshman named Monte Davidoff to help out with some of the math, and worked incessantly. "We were worried that if we didn't jump in right away, the computer revolution would happen without us," Gates tells me. Allen remembered watching Gates fall asleep in the middle of coding, his nose hitting the keyboard; after a nap, he'd squint at the screen and pick up where he left off.

By late February 1975, Allen was ready to fly out to Albuquerque, where he met with MITS's Ed Roberts, a hefty six-foot-four former Air Force officer. By then, Allen had a full beard and looked even older, all the better to negotiate a business deal. Roberts welcomed him with a three-dollar buffet dinner at a local Mexican restaurant. The next morning, Allen presented the code that he and Gates had created for the Altair computer. When Allen typed in 2+2, Roberts's computer spit back 4.

That successful outcome catapulted Gates and Allen from diehard programmers to software entrepreneurs. That fall, Gates took a leave of absence from his junior year at Harvard and moved to Albuquerque, where he and Allen signed on to create a marketplace-ready version of their prototype software. As their bleary-eyed friends back on campus pecked out research papers on typewriters, blotching out mistakes with Wite-Out, Allen and Gates leased their first office space with views of the desert valley.

The potential was clear, and it was big: Computers were about to move beyond the realm of big business and tech geeks to the consumer

Spark

market. The code Gates and Allen were writing to run the Altair was just the beginning. Soon, their software would make its way into millions of homes around the world; eventually, they would dominate the computer marketplace. In the spring of 1975, Gates and Allen set out to ignite the PC revolution.

⌒⌒

There was a time when genius was viewed as a solitary undertaking—the prowess of individual brilliance. Today, technological innovation is viewed widely as a team sport. On October 4, 2017— the very day that the Royal Swedish Academy of Sciences announced that Richard Henderson and two other scientists had won that year's Nobel Prize in Chemistry for their development of cryo-electron microscopy—the *New York Times* published a letter signed by seven eminent researchers, including Henderson himself. "As Nobel laureates, we know better than most that these landmark achievements are rarely the work of one individual," the scientists wrote. "Collaboration, across disciplines and borders, lies at the heart of scientific discovery, and without it we may never have received our individual accolades."

Critics have long charged that awarding highest honors and a gift of 10,000,000 Swedish krona ($1.1 million in U.S. dollars as of 2020) to an individual scientist ignores the teamwork inherent in any major discovery, elevating the myth of the lone genius secluded in a lab until the aha moment materializes from deep in his solitary mind. At the start of his Nobel lecture in Stockholm, Henderson graciously accepted his shared award. "But I wanted to begin," he said, "by saying that the three of us are really representatives of what's now become quite a closely knit and well-organized field." At various points in his presentation, Henderson showed images of other scientists executing critical work in the field; he concluded with a photo gallery of 21 of his own collaborators. "We are only the tip of the iceberg," he said.

Innovations rely on collaboration, whether they emerge from a physicist building on Pascal's law or a cadre of 20-somethings sharing take-out sushi while designing a new app. This is true not just in science, but also across disciplines, says Keith Sawyer, professor of education at the University of North Carolina at Chapel Hill. In his book *Group Genius,* Sawyer argues that history is littered with misconceptions about celebrated breakthroughs as solo feats, from Thomas Edison's lightbulb to T. S. Eliot's 1922 poem, *The Waste Land.* "Collaboration drives creativity because innovation always emerges from a series of sparks—never a single flash of insight," Sawyer writes. "When we collaborate, creativity unfolds across people; the sparks fly faster, and the whole is greater than the sum of its parts."

For a moment, I wondered if Sawyer was stretching too far with his conviction that even the most independent of minds are involved in some kind of collaborative effort. But then I called up photos taken at Edison's Invention Factory in Menlo Park, New Jersey, in 1880— and there he is, surrounded by a bevy of lab assistants. Sawyer also pointed me to the handwriting on the original manuscript of Eliot's *The Waste Land,* which included edits by the poet's wife, Vivienne, and his editor, Ezra Pound, whose copious jottings included "too loose," "too penty," and "too tum-pum at a stretch." "People do have ideas when they're by themselves," says Sawyer, "but the power of collaboration makes you more creative, even when you're alone."

Sawyer believes so strongly in what he calls an "invisible web of collaboration," that when we talk over the phone I can practically hear him cringe at the mention of lone genius. The United States is especially prone to this idealization, he believes, because of its roots in historical lore: the rugged pioneer who sets out to seek his fortune in

Spark

new territory. In his book, Sawyer describes a classic 1931 study by psychologist Norman Maier, who wanted to see how people conceptualize their moments of discovery.

Maier hung two ropes in a large room and asked participants to conceive different ways to tie the two ends together. The ropes were too far apart for subjects to hold onto one while walking to the other, but they were allowed to use tables, chairs, extension cords, pliers, and other items in the room to complete the task.

Maier's subjects came up with a few possible solutions, including using an extension cord to lengthen one of the strings. But the scientist was most interested in a more elegant approach: adding a weight to one of the ropes, swinging it like a pendulum, and grabbing it as it neared the other. Several participants figured this out. But most did not, so Maier provided hints, at one point casually brushing past one of the ropes and setting it in motion.

Suddenly, some of the students who had not thought of the pendulum solution realized what to do. But when asked *how* they knew what to do, they failed to mention Maier's clue. "It just dawned on me," one of the subjects said. Sawyer argues that this is an example of confabulation—a psychological term for making stuff up to fill in memory gaps—that is repeated and morphs into legend. "It's part of our national character to be receptive to stories of the solitary individual," he says.

When he was working toward his doctoral degree in the 1990s, Sawyer studied under Mihaly Csikszentmihalyi, the psychologist who pioneered the concept of "flow." In his investigations of chess players, rock climbers, dancers, composers, athletes, artists, and others, Csikszentmihalyi discovered that what makes an experience most satisfying is a state of a consciousness in which individuals become wholly absorbed by their creative pursuits—so much so that they lose track of time, feel as if they are in "effortless control," and perform at peak ability. The flow experience transpires when people engage in an activity they love, stretching their body or mind to achieve something challenging and meaningful.

Through his own stints as a pianist performing in jazz ensembles, Sawyer has experienced the bursts of inspiration that ripple through a group. Every musician's individual notes—from the pianist to the horn player—adds to a joint process of inventiveness. Drawing from Csikszentmihalyi's work, Sawyer calls this process "group flow," which he believes takes over when clusters of people come together in sync. In the case of jazz performers, he says, "that musical interaction drives the creativity forward."

This kind of interactive rumination can occur in two-person partnerships as well: Gates and Allen. The Gershwins and Wright brothers. Susan B. Anthony and Elizabeth Cady Stanton. Each boosts the other's strengths. After Anthony and Stanton met at an antislavery meeting in Seneca Falls, New York, in 1851, they joined forces to lead the women's rights movement, with Anthony serving as tactical leader and Stanton as thinker and writer. "Together, we did better work than either could alone," Stanton reflected. After Stanton's death in 1902, Anthony remarked that the greatest pleasure of their working relationship was "when she forged the thunderbolts and I fired them."

Partnerships not only bolster creativity and momentum; they are also good for our health. Studies have found that people with stronger social relationships live longer. Weak relationships, by contrast, may have ill effects as detrimental as smoking and drinking alcohol—and worse than being obese or sedentary.

It's even better if your spouse or partner is satisfied. An analysis of 4,400 couples from the University of Michigan's Health and Retirement Study revealed that when a person's partner reported greater satisfaction in life that person had a 13 percent lower risk of dying over the course of eight years, the length of the researchers' analysis.

Spark

In 2019, a playful Bill and Melinda Gates appeared on *The Late Show*'s Pop Quiz. Sitting next to each other—Bill in a trademark lavender V-neck sweater, Melinda in a leather jacket—they wrote out answers to questions posed about their relationship. They agreed on who was worse at responding to texts and had a harder time getting out of bed in the morning (Bill) and revealed each other's guilty pleasures. Bill wrote "exercise" for Melinda; she responded "playing bridge online" for Bill.

In response to the query, "What is one adjective to describe your relationship?" Melinda picked the word "fun." Bill zeroed in on another aspect. "There are a lot of words I could have used," he said.

But when he held up his board, written in his left-handed scrawl, it had just one: "partners."

⌐

Bill Gates has long stressed the importance of collaborators in his life, beginning in Albuquerque with Paul Allen. In late November 1976, the two registered the trade name Microsoft (for *micro*processors and *soft*ware) with the Office of the Secretary of the State of New Mexico and began their business alliance. As Allen saw it, he was the ideas guy; Gates homed in and made them a reality. He and Allen were interactive resources for each other, Gates later wrote. "He had a wide-ranging mind and a special talent for explaining complicated subjects in a simple way," he observed. Allen also introduced Gates to Jimi Hendrix; later, after Allen bought the Portland Trail Blazers, he took Gates to games and elucidated the rules of basketball as they watched.

In Albuquerque, Gates and Allen recruited a team of young programmers, including a couple of coding friends from Lakeside, and began building their company. They were off to an auspicious start. In their contract with Ed Roberts, Gates and Allen got a deal that weighed heavily in their favor: MITS agreed to license Microsoft's software, allowing Gates and Allen to retain ownership; Roberts also consented to use his "best efforts to license, promote, and commercialize" the program, and to give Microsoft royalties as well. In so doing, Microsoft retained the rights to license their software to competitors; as the personal computer market exploded, they attracted new clients. One of these was newly formed Apple, which signed a deal with Microsoft for its popular Apple II computer in 1977.

In this heady era of Microsoft's debut, Gates famously pushed the limits of his power. While he had an unwavering focus and sharpness of vision, he was known to be tempestuous and headstrong, especially when coders didn't live up to his expectations. He also had his eye focused on building the company's profit in a way that critics decried as ruthless and bullying. In Allen's words, "Microsoft was a high-stress environment because Bill drove others as hard as he drove himself."

Gates acknowledges this intensity today. "When I was in my 20s, I didn't believe in vacations. I worked every weekend. I was fanatical," he tells me. And he expected others to have the same obsessive commitment. "In the early days of Microsoft, I knew every employee's license plate number so I could tell when they were in their office or not." Sleep wasn't planned; it just happened. "He lived in binary states: either bursting with nervous energy on his dozen Cokes a day, or dead to the world," Allen recalled. Sometimes, he'd return to work in the morning and "see Bill's feet sticking out of his office doorway in a pair of scuffed loafers."

In his limited spare time, Gates gambled on risky diversions. He liked to slide down the balustrade in Allen's lake house; he bought a Porsche 911 and started loading up on speeding tickets. A famous mug shot shows 22-year-old Gates with moppy hair, tinted glasses, and a smile better suited to a high school yearbook. "Fortunately,

I didn't kill myself doing that," he later joked. One night, he called Allen, who scrounged up enough money—including coins on Bill's dresser—to bail him out.

Looking back, Gates says his workaholic mindset made sense at the time. "When I created Microsoft with Paul, we could see the computing industry's potential in front of us," he says. "Our mission was to have a computer on every desk and in every home. We knew that the more sophisticated the technology sector became, the more it would impact every other sector. It could improve medical research, education, the way cities are built and designed."

By the time the pair relocated Microsoft to Seattle in 1979, company sales had exceeded a million dollars. But their biggest deal was still to come. In the summer of 1980, not long after Gates hired his old Harvard classmate Steve Ballmer as Microsoft's first business manager, the company launched into negotiations with IBM. It would be a major get: Produce programming language and also an operating system—a computer's central command program—for IBM's first personal computer.

The language was easy. But rather than design an operating system from scratch, Microsoft bought one for $50,000 from a small Seattle company and modified it. Then Microsoft nailed down an agreement with IBM that echoed their advantageous deal with MITS: IBM's license for Microsoft's program, which they named MS-DOS (Microsoft Disk Operating System), would be nonexclusive, allowing Microsoft the right to sell it to other companies. And IBM could not alter it in any way to make it unique to their company. "Our goal was not to make money directly from IBM," Gates later noted, "but to profit from licensing MS-DOS to computer companies that wanted to offer machines more or less compatible with the IBM PC."

Microsoft's strategy became a pivotal factor in the company's explosive success—and Gates's unprecedented wealth. By 1985,

Microsoft had almost 1,000 employees and revenues of more than $140 million. After the company went public in March 1986, Gates earned his title as youngest self-made billionaire in history at the age of 31. By 1992, he was the richest man in America. Three years later, *Forbes* announced he'd reached the pinnacle: richest man in the world.

⌒

The IBM deal spurred Microsoft's dominance of the software market, but it also tested the relationship between Gates and Allen, whose differing work ethics often clashed. Gates's nonstop exertion took its toll on Allen, who was far more eager to take breaks for subjects that fascinated him outside the office, including space rockets.

In the spring of 1981, just as Microsoft was finalizing its IBM operating system, Allen bolted to Florida to see the maiden launch of the shuttle *Columbia,* infuriating Gates, who hadn't given him permission to leave. Allen never second-guessed watching the orange glow of the ignited rocket boosters in person. "I had seen a rare thing," he later reflected. In 1982, Allen was diagnosed with Hodgkin's lymphoma at the age of 29. One year later, with his illness in remission, he left the company.

Gates's intense focus on business left family members wondering whether he would be able to make time for anyone else. But in 1987, Gates met Melinda French, a newly hired Microsoft programmer, at a company dinner and asked her out on a date. Nine years his junior and a Duke M.B.A., Melinda had learned how to code on an early Apple desktop at Ursuline Academy, an all-girls Catholic high school in Dallas. The two loved building software and shared a conviction that it had infinite potential.

In the early days of dating, Bill suggested that Melinda switch on the green banker's light in her office to indicate that it was a good time for him to visit—a nod to Jay Gatsby's infatuation with the green light that burned at the end of Daisy's dock in Bill's favorite novel, *The Great Gatsby*. In addition to his literary allusions, Melinda liked Bill's taste in music—which included Frank Sinatra and Dionne Warwick—his curiosity and optimism, and the heart she discerned beneath the tough-minded dealmaker. And she had plenty to offer him: a keen mind, an effusive laugh, a willingness to disagree, and a competitive spirit that matched his. She beat him in their first game of Clue.

Gates's relationship with Melinda required that he begin to readjust his priorities. Until then, there had been no reason to scale back on his incessant work hours. Melinda, and the prospect of children, changed that. "When we got engaged, someone asked Bill, 'How does Melinda make you feel?'" Melinda writes in her book, *The Moment of Lift*, "and he answered, 'Amazingly, she makes me feel like getting married.'"

In the fall of 1993, Bill and Melinda celebrated their engagement by taking a trip to Africa, where they delighted in the savanna wildlife. A photograph shows them on safari looking out of a sand-colored jeep with broad smiles. Bill is wearing a dark T-shirt and wire-rimmed glasses, his hair tousled over his forehead; Melinda has on a khaki overshirt with epaulets and a scarf tied loosely around her neck. Her bangs sweep down over her eyes. Thousands of miles away from their Microsoft offices in Seattle, they appear relaxed and warmed by the sun. "I will never forget," Melinda later wrote, "how small I felt under that huge, stretching sky, surrounded by nature."

As awed as they were by the panorama and the animals, the people of Africa and the conditions in which they lived affected Bill and Melinda most profoundly. "I remember peering out a car window at a long line of women walking down the road with big jerricans of water on their heads. How far away do these women live? we wondered. Who's watching their children while they're away?" Gates later recalled. "That was the beginning of our education in the problems of the world's poorest people."

At the end of their journey, the couple took a stroll on the beach in Zanzibar, an island off the coast of mainland Tanzania. Back home, they had talked about their intent to give back to society but had not yet come up with a plan. Now, as they walked through the sand with the pellucid waters of the Indian Ocean stretched out before them, they asked themselves what they could do to improve the lives of men, women, and children who had few resources of their own. "The landscape was beautiful. The people were friendly. But the poverty, which we were seeing for the first time, disturbed us," Gates later said. "It also energized us."

On New Year's Day 1994, several months after returning from Africa, Bill and Melinda were married on the Hawaiian island of Lanai. In a toast to her new daughter-in-law, Mary Gates offered some advice: "Celebrate his good points and remember you don't have to love everything about him," she said. "Don't expect calm waters. Pray for courage. Keep your sense of humor."

By then a prominent civic leader and philanthropist in Seattle, Mary Gates had often asked her children: How much of your allowance are you giving to the Salvation Army at Christmas? At the end of her wedding toast, she left the young couple with a mission, quoted from the Bible: "To unto whom much is given, of him shall be much required."

It was an especially poignant moment for the family; Mary was seriously ill with breast cancer and would die just five months later—a day Gates later said was the worst of his life.

Mary's words and the example she had set in charitable work in her life reinforced Bill and Melinda's determination to help others. But neither one knew very much about the scale of human suffering around the world. In the months and years that followed, the couple began traveling to developing countries, meeting with experts, and consuming reports and news stories about economic, educational, and health disparities. Bill was distressed by the calculus of despair— the millions of people who were dying from treatable diseases back home. Melinda couldn't stop thinking about the women and babies.

Gates had always envisioned turning to philanthropy later in life, in his 50s or 60s. But the statistics of inequity pulsated in his brain. A *New York Times* article documenting hundreds of thousands of deaths from contaminated water galvanized his and Melinda's resolve. The piece, written by Nicholas Kristof in 1997, reported that 3.1 million people—most of them children—were dying of diarrhea in developing nations. "That's when the work in global health started for us," Melinda writes in her book. "We began to see how we could make an impact."

In 1999 and 2000, Bill and Melinda transferred $20 billion of their Microsoft stock—about a quarter of their net worth at the time—to form the Bill & Melinda Gates Foundation, making it the largest private charitable organization of its kind in the world. And Gates made an announcement that took many by surprise: He was stepping down as Microsoft's CEO to focus on developing software, leaving Steve Ballmer in charge of operations. It was sooner than Gates, then 44 years old, had intended. But he was drained by the government's antitrust suit against Microsoft (launched in May 1998 and settled on appeal in 2002), which had rallied critics to declare him a monopolist villain. And his priorities had shifted as he traveled in Africa and Asia and witnessed extreme poverty, poor sanitation, and lack of access to health care and lifesaving technologies. The impact on young children affected him most urgently. "No one can see a child with polio or malaria and not be moved by it," he wrote to me.

When he launched Microsoft with Allen as a brash young whiz kid, Gates was juiced by his bold faith and competitive drive. He built the software that allowed humans to communicate with computers, and customers needed his products. As a newcomer to global development, he threw himself into what he knew how to do best: read, read, read. And ask questions. How do health systems work and how can we improve them in the world's poorest places? What will it take to avoid a climate disaster? How do we deliver vaccines to remote areas? What can we do to make water cleaner? He inhaled as much data and research as he could, loading his brain with complexities.

Still, he wondered: Would anybody listen? One day in 2001, Gates found his answer. Warren Buffett had invited him to speak to a group of business leaders about his new work. As Gates talked about the potential for everything from vaccines to alleviate disease to toilets that could stop the spread of diarrhea, he could feel himself become increasingly energized. "When ideas excite me, I rock, I sway, I pace—my body turns into a metronome for my brain," he later wrote in a blog post titled "The day I knew what I wanted to do for the rest of my life." In articulating the burden of poverty and the benefit of lifesaving R&D in science and technology for others, he gave himself the confidence to move forward. "For the first time, all the facts and figures, anecdotes and analyses cohered into a story that was uplifting—even for me."

That story, which had begun as an intellectual challenge, started to consume Gates on a deeper level. He began taking on a more public and active role in global health, dedicating increasing amounts of time to his philanthropic partnership with Melinda as they raised their three small children. Saving lives became an integral part of their marriage, deepening their understanding of each other.

The two are often seen as having distinctly different approaches to problems, but Gates says that is a misperception. "People often think that I focus on numbers and Melinda focuses on people, but that isn't true about me or her," he tells me. "She has an incredible mind for data, and it's impossible for me to do this work without thinking about the people who are suffering the most."

In 2006, the Gates Foundation received a major boost when Buffett pledged the bulk of his Berkshire Hathaway stock to the organization—a gift valued at $31 billion at the time. He told Bill and Melinda to "swing for the fences." Bill was further motivated by the resolve he saw among the foundation's employees, grantees, and partners. "I realized quickly that when people hear that there are children dying of diseases we solved long ago in the rich world, they want to help," he tells me. In July 2008, Gates stepped away from his daily role at Microsoft to focus full-time on the foundation.

Since then, the organization has spent $54 billion on their initiatives, which include polio vaccination and the distribution of medicines to treat tuberculosis, malaria, and HIV. Under Bill and Melinda's joint leadership, they are tackling priorities ranging from agricultural development to gender inequality and family planning. By midsummer 2020, the foundation had committed more than $350 million toward international efforts to develop rapid diagnostics, therapies, and vaccines against COVID-19. Gates Sr., who helped navigate the foundation's vision and advocated for its projects, served as a steady presence from its inception—a partnership Bill treasured. "Nobody has influenced its values more than he did," he tells me.

Although Gates did not map out the order of his life trajectory, it makes logical sense to him now. By the age of 17, his software mind had been shaped; he possessed the youthful gumption to push himself to keep up with a new and booming industry. Learning how to track down software bugs and create new products parallels the foundation's efforts to navigate colossal challenges and innovate solutions, like designing sanitary toilets and using technology to track vaccine delivery. The zealousness of his youth continues to propel him, though he now sets aside more time to contemplate and write. This includes solo "think weeks" away from family and friends in a hideaway cottage alone with his books.

"I am still fanatical about solving some of the world's toughest challenges—global health and developmental issues, climate change, epidemic preparedness, U.S. education, Alzheimer's disease," he tells me, "but I don't work around the clock in the same way as I did when I was younger."

In 2007, 32 years after growing Microsoft from a concept to a technology empire, Gates returned to Harvard to receive an honorary Doctor of Laws degree. Dressed in a suit with a pink button-down shirt and striped tie, he stood before the graduating class in Harvard Yard and applauded them for taking a more traditional route to their degree than he had. "For my part, I'm just happy that the *Crimson* called me 'Harvard's most successful dropout,'" he said. "I guess that makes me valedictorian of my own special class. I did the best of everyone who failed."

Like anyone, Gates has experienced plenty of stumbling blocks. Some of the foundation's early efforts revealed the complexities of global health. Initially, Gates thought donating computers to poor communities would boost access to information, but a visit to Soweto made clear that technology isn't the answer when buildings don't have power outlets. An effort to wipe out a parasitic disease in India did not have the benefit he'd hoped for, in part because of the practical challenge of administering a medication over successive days in rural areas.

Gates has his own personal regrets, which range from the relatively inconsequential to the profoundly significant. He's sorry he did not study a foreign language as a child; he took Latin and Greek in high school, but would have liked to learn French or Arabic or Chinese.

He has realized that intelligence is not one-dimensional and not as important as he once thought it was. And although he famously raised alarm bells about lack of preparedness for an epidemic in a TED Talk in 2015—"We're not ready," he warned—he expressed frustration after the COVID-19 outbreak in the spring of 2020.

"I wish I had done more to call attention to the danger," he said in an interview with the *Wall Street Journal*. How much more of a focus he could have commanded before the virus emerged is unclear; he spoke to the presidential candidates in 2016, as well as other leaders in the years following. But Gates feels personal angst about what transpired. "The whole point of talking about it was that we could take action and minimize the damage," he observed.

Over the years, Gates has reached out to the people who made a difference in his life. In 2010, he flew to the bedside of Ed Roberts, who was gravely ill with pneumonia in Macon, Georgia. Even though Roberts had left the computer business decades earlier—he sold MITS a few years after developing the Altair computer and became a small-town physician—he had played a pivotal role in the launch of Microsoft. In a statement issued after Roberts's death, Gates and Allen acknowledged his role in their success. "Ed was willing to take a chance on us—two young guys interested in computers long before they were commonplace—and we have always been grateful to him."

Gates's relationship with Allen was, of course, far more complicated, but the two circled back to each other later in life. There had been plenty of rocky stretches. "Over time, though, the hard feelings faded," Allen later reflected. When he got sick again in 2009, Gates visited him regularly and they spoke honestly. "He was everything you'd want from a friend, caring and concerned," Allen wrote. After Allen's death at the age of 65 in October 2018, Gates remembered him as a brilliant technologist and philanthropist, and a man who embraced life with gusto and love. "I will miss him tremendously," Gates wrote in a tribute.

Gates's enormous wealth has given him a freedom few can envision and, at the same time, has required that he make choices on how to spend his money. He waits in line for take-out burgers—but he has splurged on some material comforts, including his 66,000-square-foot house (with its own trampoline room) and a private jet to help get him around for foundation work. In 1994, he bought one of Leonardo da Vinci's famed notebooks, the Codex Leicester, for $31 million. The artist inspires Gates, he said in an interview with *60 Minutes* in the way "that he kept pushing himself, that he found knowledge itself to be the most beautiful thing."

Bill and Melinda Gates intend to donate 95 percent of their wealth to charity, and have encouraged others to do the same. In 2010, the couple joined Warren Buffett in creating the Giving Pledge. So far, more than 200 of the world's wealthiest individuals—including Elon Musk, investor Shelby White, and Spanx founder Sara Blakely, have signed on to pledge the majority of their fortunes to solving urgent problems around the world.

Bill Gates, the 60-something computer magnate turned philanthropist, still has a lot in common with the towheaded child he once was, filled with boundless intensity and curiosity. The boy who devoured science fiction and stole out of the house at night to code computers has become the man who travels with a canvas tote bag stocked with books and stays up late to read. Hamburgers have been a lifelong constant, as have simple words like "cool" and "fun" to describe complex experiences. He loves winning in tennis and is still more comfortable thinking about zeros and ones—the numeric systems that run computers—than mingling with people.

In a series of "Ask Me Anything" chats that he launched in 2013 on the website Reddit, Gates began sharing revealing tidbits of his life with the public. He has a rule that he finishes every book he starts. U2 is one of his favorite bands. He watches shows like *Silicon Valley* and

A Million Little Things with Melinda, does the dishes, and enjoys a lot of sauce on his barbeque. He'd like to be a grandfather someday.

I asked Gates what he would like to achieve in the coming decades. He sees his own goals through the lens of his foundation's collaborations—with governments, the private sector, and other nonprofits—and he has a long and ambitious list. He hopes that polio will be eradicated, clean energy improved, and advances made on climate change, global poverty, and childhood mortality. "I hope that over the next decade or two, a lot of the issues I work on now will become largely irrelevant because the world has made so much progress on them," he wrote.

When Gates and I communicated in the spring of 2020, the world was entrenched in the early months of the coronavirus pandemic, which he described as perhaps the "biggest setback of my generation." Still, he said he felt buoyed by scientific efforts to seek medications and a vaccine, and hopeful about discoveries that would ultimately save lives. He had witnessed firsthand a revolution in the development of personal computers and had faith that he would see similar breakthroughs in other areas.

"I have always had a sense of optimism about the world and the potential for innovation and science to create miraculous advances," he tells me. "I would say my optimism has grown stronger over the years."

CHAPTER 5

Isaac Newton

The Physicist (1642–1727)

O ne late August morning in Cambridge, England, my husband, son, and I hopped into a rented black Vauxhall Crossland X and set out to trace the most important journey of Isaac Newton's life.

Just over 350 years ago, in 1665, Newton and his fellow classmates were forced to flee the University of Cambridge to escape bubonic plague, which had killed hundreds of people in London and was threatening the inhabitants of nearby cities and towns.

Newton was 22 years old when he left that June day for his childhood home of Woolsthorpe Manor, 60 miles to the north. The journey along what was then the Great North Road required three days and two nights of travel by horse and carriage. It was long, but not overly arduous. Today, the trip takes just over an hour on the A14 to the A1. The land is flat and alive with greenery, and the vistas filled with farmland. Even now, travelers stop at local pubs for a meal in quaint centuries-old towns along the route.

Although he could not have known it when he set out from Cambridge, Newton's interlude at Woolsthorpe—a quiet place away from the demands of university life—would stimulate his fertile mind and lay the foundations for his groundbreaking work on calculus, the spectrum of white light and color, and the laws of gravity.

As we turned onto the narrow lane leading to the manor, I felt as if we had entered a sanctuary. Set in the Lincolnshire countryside, surrounded by apple orchards, meadows, valleys, cornfields, and freshwater springs and rivulets, the place defies the bustle of everyday life. Birds flit about, breezes ramble through grasses.

The 18 months that Newton spent at Woolsthorpe Manor—comprising two long visits between the summer of 1665 and the spring of 1667—arose at an opportune moment. The third decade of life has an impressive track record among scientific greats. Newton, then in his early 20s, was old enough to explore and untangle complex subjects, but young enough to flex his mind in ways that might have proved more challenging later on.

Newton himself acknowledged the importance of youth during his annus mirabilis or "year of wonders," as it came to be known: He was in the right place at the right age. "All this was in the two plague years of 1665 and 1666," Newton wrote when he was in his early 70s, "for in those days I was in the prime of my age for invention, and minded mathematics and philosophy more than at any time since."

Newton the polymath was a complex man—brilliant, ambitious, reclusive, distrusting, and professionally vengeful. But the arc of his life was relatively uncomplicated. He never married, focusing instead on his solitary intellectual pursuits. His journey was linear, and his professional achievements sequential.

Newton's interest in the natural world emerged when he was a child, and captivated him throughout his school years, during his months in Woolsthorpe, and in his career as a university lecturer. At 27, he became a professor of mathematics, and in his mid-40s, published the results of his research in his landmark work, *Principia Mathematica,* for which he will always be known.

Standing on a pebbly pathway outside Newton's childhood home, Woolsthorpe's steward, Jennie Johns, directs my eye to the second floor: the location of the scientist's bedroom, where he spent much of his time working.

"We say the world changed here," Johns tells me, "because that's exactly what happened."

⌐⌐

Isaac Newton was born prematurely at Woolsthorpe Manor in the early morning hours of Christmas Day 1642, according to the Julian calendar England used at the time—the end of the very same year that 77-year-old Galileo Galilei died in the hilly outskirts of Florence.

Like Picasso, who claimed he was revived by a puff from his uncle's cigar, Isaac was not a robust baby. He was so small he fit into a quart-size pot, the story goes, and so weak he needed a pillow propped around his neck. Two household attendants who were sent to fetch supplies after his birth sat down on a stile along the way and took a rest, Newton later told his niece's husband, John Conduitt, "for they were sure the child would be dead before they could get back."

Isaac was named for his father, who had died several months before his son's birth, leaving the baby in the care of his mother, Hannah Ayscough. The Newton and Ayscough lineages diverged when it came to schooling. Isaac's paternal family descended from farmers, and it is likely that Isaac Newton the elder did not know how to read or write. The Ayscoughs, by contrast, were wealthier, and at least a few family members had received a decent education—including Isaac's maternal uncle, William, a clergyman who studied at Cambridge.

Throughout his life, Newton was a solitary figure prone to emotional turmoil and depression—qualities that some biographers have linked to the parental abandonment he suffered in early childhood. He never knew his father, and when he was just three years old, his mother married a significantly older man, 63-year-old Reverend Barnabas Smith. She moved to her new husband's nearby village, leaving her newly orphaned son at Woolsthorpe Manor to be raised by his maternal grandparents.

Contemporary accounts presume that Hannah deserted her young child and psychologically injured him in the process. But there is more to the story, Johns tells me. Hannah almost certainly remarried and relocated for financial security. And although her husband may have insisted that Hannah leave her son behind, it is also likely that Hannah felt compelled to do so, given Isaac's place in the Newton pedigree.

The Woolsthorpe estate had been in the Newton family since Queen Elizabeth's reign in the 16th century and now that Hannah had remarried, young Isaac—his father's only son—would have been the only legitimate heir to the manor. With England then in the midst of a civil war, his mother likely thought it imperative that Isaac remain in the house for the sake of his future inheritance.

Whatever the case, Isaac was left in the care of his Ayscough grandparents while his mother and her new husband created their own family, with three young children. Little is known about Newton's childhood during these early years, other than that his grandmother Margery was in charge of raising him. Newton later made it clear that he was no fan of the predicament he'd been left in. When he was 19, he recorded 57 personal sins he had committed in his lifetime, which included squirting water and making pies on the Sabbath; having unclean thoughts, words, actions, and dreams; robbing his mother's box of plums and sugar; and threatening to burn down the house his mother and stepfather lived in—with the two of them in it.

When he was 10 years old, Isaac's mother reappeared in his life. Widowed for a second time, after her second husband died at the age of 70, she moved back to Woolsthorpe Manor with Newton's half siblings, Mary, Benjamin, and little Hannah. It is unclear how close Isaac and his mother grew during this interval; he would later correspond with her and tend to her at her deathbed. But when he was 12 years old, she sent him to study and board at the all-boys King's School in the town of Grantham, about eight miles up the road.

In Grantham, Isaac lodged with the town's local apothecary, William Clarke, whose medicinal remedies likely influenced Isaac's burgeoning enthusiasm for gathering herbs. His interests in the world around him were vast and seemed to churn in his mind. In the garret room where he slept, Isaac covered the walls with drawings of birds and men, ships, plants, and the contours of circles and triangles.

Small and preoccupied with his own thoughts, Isaac wasn't particularly popular with his peers—especially not the boys, including Clarke's two stepsons, Eduard and Arthur. In his list of sins, Isaac admitted to stealing cherry cobs (a kind of sweet roll) from Eduard (and "denying that I did so") and getting into a fight with Arthur.

Isaac's intellect caused him strife at school, where his required subjects included philosophy, Latin, a bit of Hebrew and Greek, the Scriptures, and some elementary mathematics. He was initially negligent in his studies, Conduitt later said, and at one point ranked near the bottom of his class. A run-in with another student, who kicked him in the stomach, provoked Isaac to fight back—his rival rubbed his nose against a wall and Isaac shoved his combatant's face into the side of the church—and motivated him to want to outperform his classmates at every turn. The other students were "not very affectionate toward him," his biographer William Stukeley wrote in 1752. "He was commonly too cunning for them in every thing."

Isaac much preferred the company of girls, chiefly Catherine, Eduard and Arthur's sister, who enjoyed spending time with him outside of school. Years later, Catherine would remember him as "a sober, silent, thinking lad." Rather than engage in games with the boys after class hours, Isaac stayed close to home, where he delighted Catherine and her friends by making small tables and cupboards they could use to hold their dolls and trinkets.

He was a self-taught doer, dabbling in poetry and crafting models of a water clock and a working sundial. After watching workers build a windmill in town, Isaac created a smaller version of his own, which he augmented with animal power in the form of a mouse. He even attached lanterns to the tails of kites, frightening the townspeople who believed the swirls of luminescence were comets approaching in the darkness.

At a young age, Newton realized that experimentation was critical to scientific knowledge. In one of his earliest investigations as a student, 16-year-old Isaac figured out how to calculate the force of a strong gale by jumping into the direction the air was blowing, jumping in the opposite direction, and then comparing these measurements to a normal jump on a tranquil day. These kinds of analyses engaged his mind and impressed King's headmaster, John Stokes. By the time he graduated from school, Stokes had deemed Isaac his most esteemed student.

Throughout Newton's childhood, political peril surged. The English Civil Wars between Royalists and opposing groups began the year he was born and lasted until 1651, the year he turned nine. Along the way, King Charles I was beheaded and the monarchy overthrown. Not long before Isaac entered school in Grantham, the Puritan Oliver Cromwell had taken over as "Lord Protector" of England, Scotland, and Ireland—a controversial rule that lasted until 1660, when the monarchy was restored under Charles II.

But none of this concerned young Isaac, whose mind was grounded in the earth and tipped to the skies—to nature, not the state of national affairs. He was a born philosopher, wrote Stukeley, with a boundless curiosity that seeded his mind and layered it with wisdom.

"Thus early did that fruitful, that sagacious, that immense genius show its self," Stukeley wrote, "which since has fill'd, rather comprehended the universe!"

Throughout history, children have followed in their parents' professional footsteps, pursuing passions that fall naturally into their lives. Irène Joliot-Curie, daughter of physicists and Nobel laureates Marie and Pierre Curie, became a chemist and won her own Nobel Prize. Jazz great Duke Ellington's mother and father played the piano and signed their son up for lessons when he was seven years old. Picasso's father was an artist; Shirley Temple's mother an aspiring dancer; Yo-Yo Ma's mother, a singer, and his father, a violinist and music teacher.

Others choose entirely different paths—in some cases breaking out from troubled or unimaginable circumstances. Scientist George Washington Carver was born into enslavement on a Missouri farm in the 1860s, and battled the inhumanity of racism. But he was driven to learn, and with help from a couple who supported him, went on to study science and became director of the agricultural department at the Tuskegee Institute under Booker T. Washington. J. K. Rowling

Spark

wrote her first story when she was about six years old. But before penning her mega-successful *Harry Potter* series in the 1990s, Rowling struggled as a single mother receiving public benefits and battling severe depression.

And then there are those whose deep interest or profound talent seems to emerge from a distant galaxy, wholly disparate from ancestral tradition. Leonardo da Vinci, born in 1452, was expected to become a notary like his father. Instead, he unraveled the enigmas of mollusk shells and aortic valves in his exquisite chalk renderings. Martha Graham, nearly 500 years later, pushed ahead with her passion for movement, even though her psychiatrist father and her mother did not approve. In 1926, she founded the Martha Graham Dance Company and revolutionized the art form.

Newton's professional journey was supposed to mirror that of his family; the son and grandson of farmers, he was destined to work the land and oversee the manor. His mother knew this and made every effort to steer him toward it. When he was 16 years old, she summoned him home from Grantham so he could learn to run the family estate.

But young Isaac had no interest in the path laid out for him. He did not want to tend to crops and animals, nor was he any good at it; almost immediately, he neglected his responsibilities. Constantly sidetracked by his mind, he let the sheep wander away and the cattle trample the corn. He preferred to read books under a hedge or build a waterwheel and watch the glittering spray, noted his biographer David Brewster in his 1855 account of the physicist's life. Records show Newton was fined for letting his pigs roam without supervision and his fences fall into disrepair.

Clearly, he was not cut out for the family vocation. With a brilliant intellect, but no professional path mapped out, Stokes and Isaac's uncle William intervened and succeeded in persuading Hannah to free Isaac from the manor to go back to school. It would be "a great loss to the world, as well as a vain attempt," Stokes told her, "to bury so promising a genius in rustic employment." (His mother's servants, for their part, were thrilled to see the boy leave; in their minds, Stukeley noted, he was foolish and "fit for nothing but the 'Versity.'")

In June 1661, after completing his studies in Grantham, Newton enrolled at Trinity College, Cambridge, at the age of 18. From his earliest days on campus, it became clear that he lived most comfortably in his mind. Like fellow physicists Henry Cavendish and Albert Einstein, born, respectively, one and two centuries later, Newton had an extraordinary capacity to ignore the tumult around him, withdraw into his thoughts, and focus solely on the urgent scientific conundrums he wanted to solve.

This kind of self-imposed intellectual quarantine defined the young scientist's years in Cambridge, where he led a largely alienated existence. He entered university as a "subsizar," requiring him to earn his tuition by acting as a servant to wealthier classmates. Subsizar chores included fetching bread and beer for students and even emptying their chamber pots. It is not clear why Newton held this status—his mother had acquired wealth in an inheritance from her second husband—but whatever the reason it could not have helped Newton curry favor among his classmates.

Newton did manage to find a companion in another student named John Wickins, with whom he shared a room. But he had little patience for either the social aspects of university or the required curriculum, choosing instead to focus almost solely on mathematics and science (or "natural philosophy" as it was called in his day). His mind never stopped.

When bubonic plague hit London in 1665, Newton's only option was to flee. Prior outbreaks of the pestilence, or "the death," as it was called, had killed tens of thousands in Europe, and the university, approximately 60 miles north of the city, was taking no chances. The onslaught turned out to be especially virulent: Over some 18 months, about 100,000 people would lose their lives—roughly a quarter of London's population. Classes were canceled and gates closed.

With the alarm sounding, Newton gathered up his books and headed home. From there, his future unfolded: University had grounded him in learning; the plague propelled him to ponder and experiment; revelations begot scientific principles; and scientific principles reordered the universe.

Today, Woolsthorpe Manor is preserved as a living museum. Visitors can glimpse an original 1712 portrait of the scientist by Sir James Thornhill; a 1729 copy of *Principia,* Newton's treatise laying out the foundations of classical mechanics; and dim outlines of drawings scratched onto the walls. One depicts a church, possibly St. Wulfram's in Grantham, and two others are in the shape of a windmill, one of Newton's greatest fascinations as a child.

The most meaningful spot by far is Newton's second-floor bedroom, where he worked out his scientific meanderings in the Lincolnshire limestone, using charcoal left by dying embers in the fire. In this space, Jennie Johns tells me, Newton performed calculus (or fluxions, as it was known then), considered gravity, and conducted his earliest light experiments.

Johns points to a small window where Newton had created a miniscule hole in the shutters, allowing a thin stream of light to enter the darkened room. As the light made its way through a glass prism, he observed that a spectrum of rainbow colors appeared on the opposite wall of the room, 22 feet away. The result was not unexpected; children were given prisms as toys and delighted in the colors that leapt from the glass in the sunlight. But how?

At the time, conventional wisdom held that a prism created colors out of white light. Newton wasn't convinced. One day, he added a second prism to his experiment and came up with a critical new finding: A single beam of color that emerged from the first prism stayed true to its original hue after traveling through the second. The prism had not manufactured colors out of white light; instead, as it traveled through the sharp edges of the glass, the light was bent or refracted, allowing its individual components to show. Newton's observation was groundbreaking: White light was actually *made up* of colors.

"We get people in tears in Isaac's room" as visitors soak in the ambience and reality of what happened here centuries earlier, Johns says. "There's an emotional connection." She guides us toward the fireplace and a second window, which is slightly ajar, letting in thick swatches of light that illuminate the bedroom floor. The view outside the glass panes is majestic: a moody sky, a vibrant garden, and—encircled in a woven twig fence—a single apple tree draped in green and red fruits.

Outside, we meet manor guide Mike Kempster, a retired physics teacher who concedes that many of the 48,000 people who visit Woolsthorpe Manor every year presume that Newton's apple tree is a myth. But Kempster, an enthusiastic docent, sees this as an affront to the scientist's legacy. Dressed in sand-colored jacket, gray pants, and glasses turned dark by the sun, he lays out a series of interconnected dots that he and manor historians believe offer a clear conclusion. "We've got a tree here," Kempster says as he gestures toward it, "and we have stories."

Reaching for a piece of paper in his jacket pocket, Kempster explains that toward the end of his life, Newton told several biographers that he began thinking about the laws of gravity while sitting in his garden—and he made sure people knew that he had done so. One of these listeners, Stukeley, documented the account, which

Spark

Newton told him on the evening of April 15, 1726, about a year before the great scientist died:

> after dinner, the weather being warm, we went into the garden, & drank thea under the shade of some appletrees, only he, & myself. amidst other discourse, he told me, he was just in the same situation, as when formerly, the notion of gravitation came into his mind.

> why should that apple always descend perpendicularly to the ground, thought he to him self: occasion'd by the fall of an apple, as he sat in a contemplative mood: why should it not go sideways, or upwards? but constantly to the earths centre? assuredly, the reason is, that the earth draws it. there must be a drawing power in matter. & the sum of the drawing power in the matter of the earth must be in the earths center, not in any side of the earth. therefore dos this apple fall perpendicularly, or toward the center. if matter thus draws matter; it must be in proportion of its quantity. therefore the apple draws the earth, as well as the earth draws the apple.

But how can anyone be sure that *this* tree—a Flower of Kent variety with a gnarly trunk and flavorless fruits—is really the one, I ask Kempster. After all, the manor was some 200 acres in size—big enough to grow dozens of trees of different varieties. Kempster replies that Newton described himself sitting in the garden. Moreover, this tree was also known to a man named Edmund Turnor, who bought Newton's house some 80 years after his death. After falling over in a storm, the "Newton tree," as it had become known by then, was sketched by Turnor's brother, Charles, whose 1820 drawing shows both the tree and the house situated exactly as they exist today.

Isaac Newton

"If we come toward the tree," Kempster says waving me over, "lying flat on the ground, we can see the old trunk"—evidence of when the tree blew down in a giant storm in the early 19th century—and the place where it rerooted and grew again. And there's one final piece of evidence, Kempster tells me: Arborists have studied the tree and recently dated it as 400 years old by counting its trunk rings and examining its roots.

"So we're convinced," says Kempster, "that *this* is the tree!"

⌐⌐

Great discoveries are often mythologized as "eureka" moments— surges of genius that burst like firecrackers on the Fourth of July. Newton's observation that evening in the garden may have led to his renowned hypothesis: that the same gravity responsible for pulling apples to the ground also held the moon in orbit around Earth. But his theory didn't come out of nowhere—and nowhere in the account his friend Stukeley documented or anywhere else does Newton mention that the apple hit him on the head.

By then, Newton had filled his mind with the work of his forebears: Aristotle and scientists who followed. He knew the writings of René Descartes and Galileo, who had rejected the Aristotelian theory that Earth sits motionless at the center of the universe. He stayed outside at night staring at patterns in the skies, notes biographer James Gleick, and kept notebooks filled with his observations. He thought a great deal about the natural world.

Moments of awareness, like Newton's flash of clarity in his garden, often arise at unexpected times—in a dream, in the shower, on a walk—and after periods of studied contemplation. It takes time to reach an epiphany. When researching a problem, we seek knowledge consciously. We read books or try new things. But insights happen when a solution is processed unconsciously and then leaps out, surprising us in a welcome way.

Creative minds are messy, says Columbia University cognitive psychologist Scott Barry Kaufman, filled with layers of disorganized knowledge that require time to ripen. "Great ideas don't tend to come when you're narrowly focusing on them," he tells me. "You need to gather information first and get the fruits of material in your head before you can let your mind do its work."

History supports this theory. Storied accounts of the great Greek mathematician Archimedes figuring out how to measure volume while sitting in a bath often fail to recognize the knowledge he had already acquired. As the legend goes, he leaped out of the tub and shouted "Eureka!" (Old Greek for "I found it!" and thus the origin of "eureka moments") while streaking across the town of Syracuse in ancient Sicily. Did it really happen? Maybe.

But what's more important is that Archimedes was primed for discovery. Because he was trained in math, physics, astronomy, and engineering, the region's king, Hieron II, had turned to him, and not to some random lackey, to figure out how much gold was in his crown—which led to his celebrated breakthrough.

The world's most creative minds have experienced this phenomenon. Henry David Thoreau documented the necessity of wandering to unlock his ideas. "How vain it is to sit down to write when you have not stood up to live!" he wrote in his journal. "Methinks that the moment my legs begin to move, my thoughts begin to flow."

Julie Taymor, the theater director and creative force behind *The Lion King,* says that some of her most inspired insights surface after rest. "A lot of my strangest ideas come from early morning sleep, and it's really an incredible moment," she said during a forum on the enigma of genius at the World Science Festival in New York City. "I get up and the thing has become very clear very fast. It's not deliberately done."

Isaac Newton

When ideas begin to flow, it's important to be ready. Russian chemist Dmitri Mendeleev assembled his table of periodic elements in his sleep. "I saw in a dream a table where all the elements fell into place as required," he wrote in his journal. "Awakening, I immediately wrote it down on a piece of paper." Oliver Sacks, the neurologist and writer, started keeping journals when he was 14 years old so he could jot down ideas as they transpired. "I always keep a notebook by my bedside, for dreams as well as nighttime thoughts, and I try to have one by the swimming pool or the lakeside or the seashore," he wrote in his autobiography.

Cognitive neuroscientists are now making fascinating inroads into how insights might work at the cortical level. Brain imaging studies by researchers John Kounios and Mark Beeman show that an aha moment originates in the right cerebral hemisphere of the brain, just above the ear, and is preceded by a "brain blink"—a quick period in which the brain shifts focus inward, away from external distractions to one's own internal thoughts. The scientists compare this to what happens when someone is asked a question and needs time to think before answering. Often, without even realizing it, the person will shut her eyes to concentrate fully on what she wants to say—shutting out visual diversions like the interviewer's colorful scarf or the sight of someone walking a dog outside the window. A brain blink "allows one's attention to find the new idea and jolt it into consciousness," Kounios and Beeman write in their book, *The Eureka Factor*. "After all, you can't see the stars when the sun is out."

Interestingly, the researchers also discovered that state of mind matters: Being in a relaxed and positive mood allows us to open our minds, think more broadly, and make connections between seemingly different things—a key feature of creativity. This fuels insights more successfully than worrying, which constricts the thought process and gets in the way.

These discoveries might help explain why Newton's insights surged during the quiet months he spent at home, away from the tumult of university life. In a peaceful place without distraction, he could let his brain churn at its own pace—a process of intellectual awakening he understood to be true. "I keep the subject constantly before me," he said, "and wait till the first dawnings open little by little into the full light."

Newton's studies at Woolsthorpe Manor mark a pivotal moment in the scientist's life when he was in his early 20s—the "prime of my age for invention," as he described it.

Why do some people come to their ideas early in adulthood and others much later? Srinivasa Ramanujan, the Indian mathematician, began filling notebooks with calculations when he was a teenager. By the time of his death in 1920, at age 32, he had published proofs for complex mathematical equations that had confounded more experienced scholars.

Ada Lovelace, daughter of poet Lord Byron, launched a correspondence with her mentor Charles Babbage about her fascination with mathematics when she was 17. When Babbage designed what he called his "analytical engine," an early computer, Lovelace embraced the concept. In 1843, the year she turned 28, Lovelace published extensive notes chronicling how to make it work, including an algorithm often referred to as the world's first computer program.

The relationship between age and achievement has long been a subject of fascination for researchers. In the 1950s, a psychologist named Harvey Lehman charted the accomplishments of some 30,000 individuals and found that scientists and mathematicians produced their breakthrough works in their 20s and early 30s. Einstein, who was 26 when he came up with his theory of special relativity, famously said, "A person who has not made his great contribution to science before the age of 30 will never do so." Painters tended to peak later, in their 30s, Lehman noted, and novelists produced their most consequential writing in their 40s.

But here's where things get interesting. In a fascinating study published in 2016, researchers led by Northeastern University's Albert-László Barabási examined the careers of almost 3,000 physicists whose work dated back to 1893. The team turned the long-standing assumption that age was the driving factor on its head. Although many of the scientists did produce big hits early in life, a combination of other elements mattered more than chronology: motivation, IQ, openness to new ideas, good old-fashioned luck, and most important, productivity. "Success can come at any time," Barabási said in a TEDx Talk. "It could be your very first or very last paper of your career." What matters is that you never quit.

Newton was prolific in his work, both before and after his respites at Woolsthorpe Manor. When he was 24, the young scientist returned to Trinity College, where he continued to labor diligently on his many subjects, even while neglecting his most basic needs. His shoes were worn at the heels, his hair uncombed. "He stayed in his chamber for days at a time, careless of meals, working by candlelight," notes biographer Gleick, and became so absorbed by his independent studies and experiments that he often didn't sleep.

Newton's peers left him alone, unsure what to make of their silent compatriot. No one dared to step on the diagrams he scratched out with a stick in the gravel of the college walkways. A student who worked with him later observed: "What his aim might be, I was not able to penetrate into. But his pains, his diligence at those set times, made me think he aimed at something beyond the reach of human art and industry."

Where the ancient thinkers, led by Aristotle, reasoned out their hypotheses through deduction, Newton relied on rigorous experimentation and willful logic, which required time and repetition. He designed a six-inch telescope using polished metal mirrors—more powerful than conventional glass lenses—to study the skies. He stared at the sun and stuck a bodkin (a long sewing needle) "betwixt my eye and ye bone as near to the backside of my eye as I could" to analyze how the curvature of the retina affected perception of light and color.

Newton launched his ambitious studies at a decisive moment in the history of ideas—a time when the field was vast and open for exploration. Today, science is practiced in narrow disciplines, requiring researchers to specialize rather than journey between subjects. The trend toward a tapered focus makes sense: The more we learn about the vast expanse of Earth and its relationship to the universe—from coral reefs to comets—the more complex the science becomes.

But specialization can also reduce the opportunity to deviate to an intellectual whim. In the era in which he lived, Newton would have been known as a natural philosopher—a title that has at its core the pursuit of wisdom or enlightenment. As such, he had the freedom to be led by his inquisitive mind. He could veer from light to gravity and back, allowing him to make connections and see the bigger whole.

This broad scope of knowledge, along with scientific tools that aided his research, allowed Newton to catapult science into the modern era. He devised rules that explained away the centuries-old convictions of ancient thinkers. No, Earth wasn't an unruly place where the gods controlled strikes of thunder or quakes. It was a place governed by systematic and mathematical laws—a planet that operated within the fabric of the universe.

Over the two decades following his interludes at Woolsthorpe Manor, Newton continued working on his theories. He had not figured it all out during the plague—far from it. Like an artist, he had created preliminary sketches that needed to be tweaked, colored, and polished.

Newton's rigorous and ceaseless pursuit of knowledge, as both a Trinity student and professor, culminated in the publication of his magnum opus, *Principia,* in 1687—two decades after his annus mirabilis. In this three-book work (the second and third editions were published in 1713 and 1726), Newton laid out his findings on gravitation and its role in the solar system, and presented his three laws of motion: An object at rest remains at rest, unless disturbed by an outside force, and an object in motion remains in motion; when an outside force acts on an object, it accelerates the speed and direction of the object; and for every action, there is an equal and opposite reaction.

The first edition alone, which spanned more than 500 pages, was so immense and complex that even the most brilliant minds had trouble contemplating the material. As Newton passed by one day, a university student was said to have remarked, "There goes the man who has writt a book that neither he nor anyone else understands."

The publication of *Principia* catapulted the reclusive Newton, who by then was 44 years old with a head of gray hair, into the public eye, exposing the personality traits he had shrouded in his prior existence. In midlife, Newton emerged as a human being vexed by very human behaviors.

Suddenly, his work was out in the open and subject to attack by rival scientists. Suspicious, distrustful, and extremely sensitive to criticism, he battled with German mathematician and philosopher Gottfried Leibniz, who published his version of calculus three years prior to Newton and wanted credit. He also sparred viciously with

Spark

British astronomer John Flamsteed, whose work Newton relied on to establish his theory on the motion of the moon. Flamsteed charged Newton with not sufficiently recognizing his contributions. The scientist was unkind and arrogant, Flamsteed charged, and reacted impatiently.

One of his greatest nemeses was British physicist Robert Hooke, who criticized Newton's theory of light, claimed to have discovered gravity, and accused him of committing plagiarism. In one of many angry letters exchanged between them, Newton penned his now famous maxim: "If I have seen further it is by standing on the shoulders of giants." Although generally perceived today as a modest Newton acknowledging the importance of his forebears, some biographers believe that the sentence was instead meant as an insult to Hooke, who was small in stature because of a curvature in his spine.

The genius mind is often plagued by chaos. Newton's relentless work and fierce feuding took its toll. After the publication of his landmark work, he suffered what has been called a "short-lived lunacy" or a "post-*Principia* depression." The first signs of malcontent—loss of appetite, withdrawal, moodiness, paranoia, delusions—appeared in 1692, as he approached his 50th birthday, and peaked the following year. In September 1693, he wrote to a colleague saying, "I am extremely troubled at the embroilment I am in, and have neither ate nor slept well this twelve month, nor have my former consistency of mind."

What caused this midlife breakdown? Biographers have proposed numerous diagnoses, including grief over the death of his mother, manic depression, or simply exhaustion after the publication of *Principia*. Modern science has also weighed in: mercury poisoning.

The notion sprang from a discovery in 1936, when British economist John Maynard Keynes acquired a trove of Newton's private papers, which revealed an astonishing and undisclosed truth about the scientist: He was passionate about alchemy, a subject on which he had written more than a million words.

In addition to all his calculations and theorizing, Newton had also spent an inordinate amount of time huddled in his laboratory, where he conducted several hundred experiments with sulfur, nitric acid, copper, and mercury. Along the way, he not only inhaled mercury vapor but also tasted the compound on at least 108 occasions, exposing himself to toxicity that can cause an array of symptoms, including weakness and pain as well as anxiety, insomnia, depression, and memory loss. In 1979, two researchers tested four samples of hair deemed to be Newton's and found heavy levels of metals—especially lead and mercury.

Whatever caused Newton's illness, the idea that the rational scientist—the man who had rejected ancient teachings for logic—had been quietly melting metals and mixing potions for years stunned Keynes and his contemporaries. Alchemy may have offered Newton a window into the natural world as a living and spiritual organism. But before morphing into modern-day chemistry, it was viewed as a kind of witchcraft. For Keynes, a new image of the scientist emerged: "Newton was not the first of the age of reason, he was the last of the magicians."

As he got older, Newton grew more comfortable with engaging his mathematical and scientific prowess outside the halls of academia. In 1696, at the age of 53, the scientist shifted careers and moved to London, where he became warden and then master of the Royal Mint.

At the time, the reliability of England's money was in jeopardy, with one in every 10 circulating coins believed to be fake. Newton seized the power handed to him and tracked down counterfeiters with

Spark

gleeful aggression, even sending one infamous offender to the gallows. Ever the exacting scientist, he oversaw the production of a series of new and highly standardized British coins with scrupulous attention to quality and accuracy.

Although he did not take much interest in the entertainment London had to offer—an opera tired him out in the second act and he left early—he embraced his prominent role in city life. The brooding scientist who had previously sequestered himself in his laboratory took pride in showing off his accomplishments. Once oblivious to how he looked, he now cultivated his image, commissioning portraits that depicted him in formal poses. As he aged, his narrow face grew heavier and he often wore a signature wig.

In 1703, after the death of his archrival Robert Hooke, Newton was appointed president of the Royal Society, Britain's esteemed organization of scientists. One year later, he published his second great work, *Opticks*. The volume drew on the prism research he had conducted decades earlier in his bedroom at Woolsthorpe Manor, along with investigations he had carried out in the years that followed. In it, Newton set out the principles of refraction. Although he incorrectly argued in favor of what was then known as the corpuscular theory of light, which stated that light is made of tiny particles, he also documented the concept of light radiating in waves, a notion that he could not reconcile with his understanding at the time, but that would be later proven correct.

For the final three decades of his life, Newton remained in London, where he earned a good salary and touted a new title, Sir Isaac Newton, after Queen Anne knighted him in 1705. He filled his home on Jermyn Street with a considerable collection of books and chose drapes, cushions, and a settee in his favored color, crimson. Biographers suggest he remained single and celibate throughout his life, but he was not entirely alone. For many years, his niece, Catherine

Barton, the daughter of his half sister, Hannah, lived with him in his London home.

During his tenure as president of the Royal Society, a position he held until his death, Newton continued researching his theories and oversaw new editions of his work. During this time, old rivalries with Flamsteed and Leibniz erupted anew. Flamsteed claimed Newton was corrupting the publication of an astronomical catalog he had written and accused Newton of calling him "all the ill names" he could think of; Leibniz and Newton, meanwhile, got into a renewed feud over who invented calculus, with both of them accusing the other of stealing his research.

In 1722, when he was 79 years old, Newton began suffering from kidney stones, though he continued to work and correspond with scientists. Just one month before he died, on March 20, 1727, he oversaw his last meeting at the Royal Society. The 84-year-old left no will and on his deathbed, refused the church sacrament.

Although Newton never traveled more than about 100 miles from his home and, as Gleick points out, likely never saw the ocean, his probing mind explored great distances through the heavens and the seas. The scientist's legacy is as vast as the universe he studied, explaining the vibrant colors in a rainbow, the motion of the planets, the calculations of a rocket to the moon, and how the most ordinary things work in our lives today—like traveling on trains and diving into swimming pools.

At the 200th anniversary of Newton's death in 1927, the physicist Paul R. Heyl lauded the scientist for his ability to connect and make sense of his innumerable observations. "Newton stood head and shoulders above his contemporaries because he had vision, a broad mental grasp, a good sense of perspective," Heyl wrote. "He brought order out of chaos; he had constructive talent; he was a builder, not only a collector of building materials." Had Newton been required to fulfill his family destiny as landowner, his intellectual fortitude might have been squandered. Instead, as a teenager and young adult, he puzzled out the world and laid bare his genius.

The first scientist to receive a state funeral in England, Isaac Newton is buried under a slab of black stone in Westminster Abbey, close to those who followed in his footsteps: Michael Faraday, Charles Darwin, and most recently, Stephen Hawking. A marble monument facing the church nave features an ornate stone carving of the scientist with his books and scientific instruments, trumpeting his legacy in regal fashion. A long way from the trees and breezes of Woolsthorpe Manor. A distant journey from the years of quiet contemplation.

"I do not know what I may appear to the world," Newton said at the end of his life, "but to myself I seem to have been only like a boy playing on the seashore, and diverting myself in now and then finding a smoother pebble or a prettier shell than ordinary, whilst the great ocean of truth lay all undiscovered before me."

CHAPTER 6

Sara Blakely

The Entrepreneur (1971–)

Since the debut of his popular podcast, *How I Built This* in 2016, NPR radio host Guy Raz has chatted it up with hundreds of entrepreneurs. He's asked Selina Tobaccowala of Evite what it felt like to lay off employees during the tech bubble burst; Jen Rubio of Away about why she decided to design luggage (her suitcase broke while running through the airport, leaving a trail of clothing behind); and Ben Cohen and Jerry Greenfield of Ben & Jerry's about the logic of opening an ice cream store in Burlington, Vermont. ("Did anybody just say, 'Um, do you guys realize that maybe seven months out of the year it's really cold there and it might not make a whole lot of sense?'" Raz asked.)

But in 2018, when Raz polled entrepreneurs at his first *How I Built This* Summit in San Francisco on who they'd most like to hear from the following year, a single name rocketed to the top: Sara Blakely, founder and CEO of Spanx. Blakely's company, which she famously launched with a pair of cropped panty hose in 2000, had made her the world's youngest self-made female billionaire.

One warm October day in 2019, Raz came through for his listeners. Pacing on stage at the Yerba Buena Center for the Arts in a white button-down shirt, blue blazer, and brown corduroy pants, he introduced Blakely to his caffeinated audience of innovators. "She had this idea, lots of people said it wouldn't work, that it was crazy, that if it was so great, why didn't other people do it before?" Raz said. "A lot of doors were slammed in her face. But she kept going. And that story involves every possible iteration of risk and doubt and intuition—and, yes, skill and luck."

With his arm outstretched, Raz invited his guest to make her entrance. "It is my incredible pleasure to welcome to the *How I Built This* Summit, the founder of Spanx, Sara Blakely!"

Blakely walked onto the stage in high-heeled pumps and a cherry red dress with black-and-white flowers. With her capacious smile and effervescence, she waved to applauding attendees and gave Raz a hug. Then, just as she was about to sit down, Blakely noticed a young woman in the audience holding up a sign that read: "Sara, Will You Marry Me?!" The words appeared neatly in all-caps black ink, except that "Marry" had been crossed out and replaced with "Mentor" in red letters—Spanx's signature color. Blakely stopped, pointed, and laughed. "Oh my gosh," she said. "I love it!"

Over the next 30 minutes, Blakely told stories about her journey— one rejection after the next, the self-doubt and fear, the boxes piled up in her apartment, the Spanx T-shirt she made for herself as a walking billboard. The attendee with the sign, Emily Kenison, was captivated. But it was an encounter that took place off-stage after the interview ended that sent her into entrepreneurial nirvana.

The founder of a start-up called Straplets, Kenison had just dashed out of the auditorium to visit a friend at the nearby San Francisco Museum of Modern Art when she spotted Blakely standing outside. She rushed over to introduce herself. "I guess I'll go down on one knee, I'll make a whole to-do about it," Kenison said as she lowered herself down onto the cement plaza.

Straplets are an accessory that slip over dress shoes, adding flair while also keeping heels solidly attached to a woman's foot—a need Kenison identified after awkwardly stumbling in front of co-workers in the black patent leather Manolo Blahniks she splurged on after graduating from law school.

Looking up at Blakely, who threw her head back with an encouraging laugh, Kenison asked: "Will you mentor me?" And then, in what can only be described as a Cinderella-esque moment, Kenison slipped her T-Straplet, made from vegan leather and decorated with gold studs, onto Blakely's shoe.

"Good idea, Emily, good idea. Belts for your shoes!" Blakely said. Once upright, 30-year-old Kenison and 48-year-old Blakely stood with arms draped over each other as Blakely turned to a handful of observers. "What do you think, people?" she asked rhetorically with a lilting pitch to her voice. "Which shoe looks better?"

"Well, listen," Blakely said as she high-fived with Kenison, "I think I have to mentor you. I don't think I have a choice!" Kenison, who had left her job as a lawyer, reached out and hugged Blakely. "Thank you so much, Sara," she said. "You're my inspiration!" A few minutes later, Kenison stopped to process what had just happened. "I can't believe it," she said. "I'm still shaking."

Sara Blakely—idol to countless young entrepreneurs like Kenison—grew up in a Florida beach town, an enterprising kid who planned to be a trial lawyer just like her father. But after bombing the LSAT twice, Blakely ditched her career plan and landed a job selling fax machines door-to-door. It was a line of work that would test her resilience and hone her nerve.

At age 25, Blakely was promoted to national sales trainer for the company. But the drudgery was wearing her down. One day, after being booted out by yet another uninterested customer, Blakely pulled her car off to the side of the road and said out loud through frustrated tears: "This is not my life. I'm in the wrong movie. Call the director! Call the producer! Cut!" Then she went home and wrote in her journal: "I'm going to invent a product that I can sell to millions of people that will make them feel good."

For a while, nothing noteworthy jumped to mind. But two years later, while getting dressed for a party, Blakely cut the feet off a pair of her panty hose and wore them under cream-colored pants. This simple act achieved two desired results: smoothing out her panty lines and leaving her feet bare for shoes and sandals. At 27, Blakely's spark had ignited, and she was ready to build her empire, as she famously says, "one butt at a time."

Blakely had no leadership experience, not much money, and a somewhat crazy idea. But with a mix of gumption and charm, she harnessed her intuition, courage, creativity, and grit to create one of the most successful global brands in history. At age 29 and then again at 35, she appeared on *Oprah;* at 41, she landed the cover of *Forbes;* and one year later, in 2013, Blakely became the first female billionaire to sign the Giving Pledge, joining Warren Buffett and Bill and Melinda Gates in a commitment to give away at least half her wealth to philanthropy.

At one point—before Spanx was even a kernel of an idea—Blakely's father suggested she find another job that would make her happier than selling fax machines. But Sara decided that would be too easy. "This voice inside of me said, 'But then I might be content,'" she told Raz in San Francisco. "I need to stay miserable, because if I stay at rock bottom and I'm this unhappy, I'm going to ensure that I am willing to take some of sort of leap that terrifies me."

⌒

Sara Blakely was born on February 27, 1971, to Ellen Ford, an artist, and John Blakely, a trial lawyer, in Clearwater, Florida. The early 1970s marked a momentous time in American history and culture. Anti-Vietnam War sentiment was at its height and President Nixon was secretly taping conversations in the Oval Office. Within months of Sara's birth, Walt Disney World opened its multimillion-dollar doors in Lake Buena Vista; Three Dog Night's "Joy to the World" topped the music charts; a lunar roving vehicle traversed the surface of the moon for the first time with the crew of Apollo 15; and gas hit 36 cents a gallon.

The coastal town of Clearwater, edged by salt marshes and white sand beaches along the Gulf of Mexico, made for an idyllic childhood. The Blakely family took full advantage of the Florida sun, fishing for trout and catching stone crabs off the dock in the bay. In an era yet to be inundated by helicopter parenting, Sara and her younger brother, Ford, played freely in the neighborhood streets, went water-skiing, and tooled around the bay with their friends in a Boston Whaler, which they also used to motor over to swim team practice.

As a young child, Sara had something of a philosopher's mind, contemplating her place in the universe. "I would always ask my mom at bedtime, 'Why am I here and what's my purpose?'" she says. Occasionally, she would even belt out her existential queries during shopping trip meltdowns. "I knew that you meant, 'Why am I here on earth?'" her mother later told her, "but everyone else would think you were talking about Kmart." Ellen says her daughter's questions always surprised her. "It would kind of blow me away," she tells me.

"I didn't quite know what to say, other than we really don't know, it's a mystery."

Sara's interrogations extended to run-of-the-mill kid matters, too. When I ask her father about this, he reels off the kinds of queries she used to pose: Why do I have to go to school? Why is there recess? Why does school start at nine o'clock instead of 10? Why can't I wear what I want? Sara also had an ability that parents both admire and dread: figuring out the exception to every rule. "She could outdebate her father," Ellen says. "It was difficult," John adds with a loving laugh. "She was a challenge all the time."

From early childhood, Sara's boundless energy was front and center. She could not sit still and had no interest in taking naps. Looking back, John says he first noticed that his daughter was unusual when she was four or five years old, and began assuming commanding positions among her bevy of friends. "She was always director of the play or captain of the game," he says. "I figured she was going to be a leader of some kind. I also knew that whatever she was going to do, it was going to be at 100 miles an hour."

Before long, Sara had assumed the role of chief entertainer in the neighborhood. Ford remembers his sister corralling kids every weekend with new games that she'd invent on the fly, including "witchy poo" (a variation on capture the flag) and "fun run" (an obstacle course). For hours at a time, the kids chased one another around, climbed trees, and swung on ropes. "All our friends would come over and see what trail Sara was going to blaze and see if we could follow her," Ford tells me. "She had an innate ability to lead."

Entrepreneurs lined both sides of the family tree. Ellen's grandfather owned a construction company but had a passion for geology and discovered oil in Michigan; her parents opened two clothing stores, one for men and one for women. One of John's grandfathers invented a rotating bread oven while the other ran a four-color printing press; in the 1950s, John's father gave up law in Beloit, Wisconsin, to manufacture a combination pressure cooker/deep fryer used to make "broasted" chicken, still a favorite in the area today.

Sara's entrepreneurial bug emerged early. In her first profit-making venture, when she was around seven, she suggested to her friend Carla that they peddle their rainy-day drawings to neighbors for five to 10 cents apiece. That success led to a "charm socks" business around the age of eight, in which she sewed five-and-dime store amulets onto frilly white ankle socks and hawked them at school and swim practice.

When I ask Blakely's mother about her daughter's resourcefulness, she chuckles with amazement. One day, she tells me, Sara built a putt-putt golf course in the playroom out of building blocks, beach towels, pots and pans, and her parents' putters and balls. "She'd use everything imaginable," her mother says. "The kids would just clamor to come over. They'd line up and they'd give her 25 cents to play."

Her daughter even invented an enterprising way to get her chores done. "She was extremely messy," Ellen says, "and her room was like a tornado." Rather than clean it up herself, she enlisted neighborhood kids—including five sisters who lived next door—to do the job for her. Same thing when her father asked her to weed the garden. Sara's recruits competed for a prize, and she got a break on menial labor.

It helped that Sara had a magnetic personality. She was funny and charming—a swimmer, soccer player, cheerleader, and tomboyish Sunshine State kid who loved roller-skating up and down Clearwater Beach in a bikini. "I think we all thought she was special," says her mother. Even the grown-ups liked having her around. When friends came over for cocktails or dinner, her mother says, "they'd say, 'Have Sara come down.'" She obliged, entertaining them with dialogue from the movie *Airplane.*

Sara seemed to have it all: a stable home, loving parents, a younger sidekick brother, friends who adored her, good looks, multiple talents, and a dream for the future. But no one is immune to life's upheavals. When she was in her mid-teens, the Blakelys separated and Sara's father moved out. At 16, she endured an unthinkable tragedy: A car hit and killed her friend as the two were riding bikes together over the Belleair Causeway.

The trauma ultimately compelled her to want to take charge of her life. "When you're 16, you feel like you're going to live forever," she says. "That changed my perspective and made me want to get on with the show, take risks, try new things, and not wait around." It was a maxim she would rely on when other catastrophes struck— including the death of her closest friend, Laura, in a horseback riding accident 15 years later.

Blakely's determined mindset as a teenager coincided with a gift from her father: a set of cassette tapes featuring the motivational speaker Wayne Dyer. Dyer encouraged his listeners to harness their internal capacity to move forward, make choices, and achieve goals without hoping others will do the work for them. We must do the molding and shaping ourselves, he preached, because "you are to your own life like a great painter to his or her creation."

Dyer's words were an awakening for Sara. She played his cassettes repeatedly at home and every time she got in her car—much to her friends' exasperation—memorizing entire sections and seizing on Dyer's idea that she could own her destiny without worrying about what anybody else thought. "It gave me such immense confidence in the possibility of my life and who I am," she says. "It made me want to take risks. It made me not want to live small."

Blakely achieved this end by tasking herself with uncomfortable challenges. One day, she dressed up and took herself out to lunch at an upscale restaurant in the tony neighborhood of Winter Park. "I walked in and said 'Table for one,' and I sat and dined alone," she says. "Knowing me, I probably ordered french fries." She didn't bring a book and cell phones weren't a thing yet, making the solo experience all the more unnerving. "My heart was pounding in my chest very big the whole time," she says. "Everyone in the restaurant was looking at me and trying to figure out why I was there and why I was dining alone."

Sometimes, Blakely chose to upend her physical appearance by wearing wigs on random days during the week. The first time she donned one at school—black with blunt bangs covering her naturally blonde hair—her friends were shocked. "All day, they'd ask, 'What's going on? Was this a dare?'" she says. For Sara, it was affirmation that if she could look different, she could stretch beyond the identity she'd assumed for herself or been assigned by others. "It would remind me to shake up my own self-definition of who I think I am," she says. "I like that to be surprising and evolving."

One night, when she was in high school, Sara's father took her out to dinner and told her it was time to learn the value of earning a dependable paycheck. You've got two weeks to find a job, he told her. She signed on as a cabana girl at an upscale Clearwater Beach resort, selling suntan lotion and setting up chairs and umbrellas for patrons. But after just a few days, Sara identified a far more enticing need among beachgoers: entertainment for active kids.

"I quit my job as a cabana girl and made a flyer that said 'Parents, enjoy the sun while your kids have fun with Sara,'" she says. Every morning, she walked up and down the beach, introduced herself to parents, told them about her program, and offered to keep their kids busy from 1 p.m. to 3 p.m. at the hotel pool. Cost: $8 per child. "She'd come home at night and sit on the floor of her messy room and count out dollar bills. She had wads of cash," her mother says. "She was so creative."

Three summers later, Sara set up an appointment with the hotel manager. Decked in a new suit from Casual Corner, she told him she wanted to expand her program along the beach and needed money to buy a minivan. "I was probably five minutes into my pitch before the guy literally turned five shades of red and said, 'What are you talking about? Who are you? How long have you been doing this kids' program on my property?'" she says. She had never sought permission, nor did she have insurance. "He told me to basically get off his property and was mortified that this had been happening."

Sara's poolside babysitting business came to a halt, but it was clear that even as a teenager, she had a knack for sensing opportunity and cultivating clients. "I love filling white space," she says. "I see the cracks in different things where this light is coming through."

Throughout her childhood, Blakely was laser-focused on following her father's career trajectory. Her early success in business? That was extracurricular fun. "I know it sounds strange," she says, "but I never thought it could be a career path. I wanted to be a lawyer my whole life." To prepare, she joined the debate team in high school so she could sharpen her speaking skills; occasionally, she skipped school to listen to her father's closing arguments in court.

There was one major hurdle, though. Sara struggled in school, especially with reading comprehension, and didn't do well on standardized tests. As we talked one day, she said she had recently uncovered her college application essay and offered to share it with me. In it, Blakely shrewdly presented herself as a defendant at trial. "All rise: Court is now in session," she began. She went on to argue that she should be judged not by her SAT scores but by her determination, curiosity, humor, empathy, and talent for debate. As evidence, she pointed to Einstein, Churchill, and Edison—three historical greats whose promise did not surge early on.

"A person's capability for great achievement or contribution to mankind cannot be judged by standardized tests," she argued. "Indeed, those persons who have a special talent, or genius, frequently measure very poorly against the standards of ordinary people."

Blakely made it to college, attending Florida State University, where she majored in legal communications. But testing continued to beleaguer her. In her senior year, she took the LSAT in preparation for applying to law school. After scoring poorly on the exam the first time, she signed up for a prep course, studied diligently, and took it again. "I did one point worse," she says. "I failed the LSAT not once, but twice."

Initially, she was devastated. "It didn't feel like it was a reflection of my intellect," Blakely says. But throughout childhood, her father had schooled his children in a simple premise: In defeat lays opportunity. "What did you fail at this week?" John Blakely would ask Sara and Ford at the dinner table. The goal was to emphasize the power of trying new things, whatever the outcome. "If you don't fail at something, you never really discover what your limits are," John tells me when I ask him about this. "You learn more from your mistakes than from your successes."

Failure became an accepted part of Blakely's life, allowing her to view disappointment in a whole new way. After losing her bid for class president repeatedly—in 6th, 8th, and 12th grades—she figured out how to view the losses as a strength, even noting in her college essay that, "I have successfully figured out three ways how not to win the class presidency."

Academics who study failure know that it is an integral part of success. This applies to every kind of endeavor and every kind of person—from a rice farmer in Indonesia to a cancer biologist seeking a cure in Tel Aviv. Indeed, many of the world's greatest minds toiled for years, with their biggest hits emerging only after many attempts. "Most articles published in the sciences are never cited by anybody," says psychologist Dean Keith Simonton. "Most compositions are not recorded. Most works of art aren't displayed."

Discerning these struggles can be beneficial, especially to school-children who believe that genius is an anointed gift, rather than earned through diligence and mistakes. Xiaodong Lin-Siegler, a professor of cognitive studies at Columbia University's Teachers College, uncovered this disconnect when she studied the impact of teaching ninth and 10th graders about the challenges that physicists Albert Einstein, Marie Curie, and Michael Faraday encountered in their lives.

In her study, Lin-Siegler split students predominantly from low-income areas in the Bronx and Harlem into three groups, and assigned each of them a differing account of the scientists' lives. One group read about their personal struggles (Einstein fleeing Nazi Germany, Curie working as a woman in a male-dominated field); a second group read about intellectual challenges (Curie and Faraday both saw early experiments fail); and a third, the control group, read a typical text-book overview of the scientists' extraordinary accomplishments.

After a six-week grading period, Lin-Siegler and her colleagues discovered that the groups that learned about the scientists' hardships showed significant improvement in their grades in science class—especially students who struggled the most academically. Not only were they motivated to do better in their own work, but the knowledge also made them feel more connected to the scientists as people. Students who read only about the scientists' triumphs, meanwhile—winning the Nobel Prize, for example—saw their grades decline. "When kids think Einstein is a genius who is different from everyone else, then they believe they will never measure up," Lin-Siegler said. "Many students don't realize that all successes require a long journey with many failures along the way."

Organizations and universities are now embracing defeat as a necessary pit stop on the road trip to triumph. At Smith College, a yearlong "Failing Well" program encouraged students to write a "failure résumé" and prompted faculty and staff to share their own stories. One professor plastered rejection letters all over his office door. In 2018, Teachers College opened an entire research center focused on studying failure across a wide variety of disciplines, with Lin-Siegler at its helm. At the launch, she talked openly about being rejected from three graduate schools—including the one that now employs her.

The goal is to encourage students to press on even when they feel most defeated and to recognize that losses can lead to wins. An intriguing study led by Northwestern University's Kellogg School of Management shows this to be true. Researchers looked at how young scientists seeking prestigious research grants from the National Institutes of Health fared between 1990 and 2005. They found 623 "near miss" applicants and 561 "narrow wins" and followed their work over the next 10 years. The near misses who kept submitting research proved to be gritty: As it turned out, they had outperformed their counterparts, publishing more hit papers with higher recognition. "The simplest implication," study author Dashun Wang told the *Daily Northwestern*, "is that what doesn't kill you makes you stronger."

For Blakely, botching the LSAT was like a helmswoman shifting her rudder—except that she had no clear destination in mind. Now a 22-year-old college graduate with an abandoned career plan, she was unsure of what she wanted to do. Too short to play Goofy at Disney World, she instead took a job buckling tourists into their seats at Epcot's now defunct World of Motion ride, which required her to spend eight hours a day in a brown polyester space suit. After a few months, she looked for something more stable and landed the fax machine job at an office supply company.

The position allowed Blakely to develop strong sales skills, which she would later exploit at Spanx. But it also forced her to withstand constant rejection. Receptionists ripped up her business cards; security escorted her out of buildings for soliciting. "It was brutal," she says. As the years ticked by and she approached her mid-20s, Blakely wanted out. "I had a sense of anxiousness about it," she tells me. "I was running out of time." The moment she pulled over on the side of the road, she came to a realization: *She* was the director and she needed to be intentional about how she planned her career.

That night, Blakely became her own life coach. "I said, 'Sara, what are you good at? You're good at sales,'" she says. But she took no pleasure in fax machines. Wouldn't it be joyful, she wondered, to sell something she had created that she was really excited about? Especially if it was an enterprise that would allow her to support vulnerable women and young female entrepreneurs. It was time to look to what has become her guide in all things personal and professional: "I asked the universe for the idea."

When I first heard Blakely talk about the universe, which she does often, it sounded unusual, even a bit far-fetched. Most businesspeople focus on numbers, logic, and profits, not the pulsing of the cosmos. I asked her to define what the word means to her, and she described it as a collective energy, a force field that draws people together and provides the precious gift of inspiration. "When you tap into it and align with it, it conspires for your greater good," she says.

Early one morning, when I couldn't sleep, it occurred to me that Blakely's universe harks back to the ancient Greek and Roman concept of an external genius attending each of us—a divine spirit that acts as muse. The writer Elizabeth Gilbert describes this magical entity as "kind of like Dobby the house elf," who lives in the walls, invisibly emerging and assisting artists with their work.

In her 2009 TED Talk, "Your Elusive Creative Genius," Gilbert bemoaned the loss of this age-old conviction, which was abandoned during the Renaissance when philosophers cast the gods aside and elevated the individual to all-powerful. "I think that was a huge error,"

Gilbert says. Asking a single person to be the "source of all divine, creative, unknowable, eternal mystery is just a smidge too much responsibility to put on one fragile, human psyche. It's like asking somebody to swallow the sun."

The muse, universe, Dobby—call it what you will—is often indeterminable. I learned this most profoundly when interviewing the jazz pianist Keith Jarrett for a *National Geographic* magazine story. Jarrett, who improvises his music in front of live audiences, told me that he finds it difficult—impossible, actually—to explain how his music takes shape on stage. When he sits down at the piano, he purposefully pushes notes out of his mind, moving his hands to keys he has no intention of playing. "I'm bypassing the brain completely. I am being pulled by a force that I can only be thankful for," he told me. His creative artistry, nurtured by decades of listening, learning, and practicing melodies, emerges when he is least in control. "It's a vast space in which I trust there will be music," he said.

Such faith can be liberating. Gilbert was ready to give up on her best-selling memoir *Eat, Pray, Love* until the moment she decided to break free from the clutch of writerly anguish and call in the creative guard. Lifting her face from her manuscript, she spoke out loud to an empty corner of her room. "'Listen, you, thing, you and I both know that if this book isn't brilliant, that is not entirely my fault, right? Because you can see that I am putting everything I have into this, I don't have any more than this," she said. "So if you want it to be better, then you've got to show up and do your part of the deal."

This kind of internal-external partnership has been essential to Blakely's success. Although self-driven and intentional about what she wants to do, she seeks the universe as confidante and creative collaborator. It doesn't always answer quickly. Blakely's two-year search for a breakthrough idea felt like the plotline of the classic children's book, *Are You My Mother?* Month after month, she says, "I was literally like,

'Are *you* my idea?'" Then, finally, one evening in 1998, her eureka moment hit. She grabbed her scissors, cut off her panty hose, and felt great in her cream-colored pants.

The journey from idea to product tested her resolve. While still selling fax machines during the day, Blakely spent nights and weekends researching hosiery and experimenting with designs. The homemade prototypes she created didn't hold up, she couldn't afford a lawyer to write her patent, and manufacturers refused to take her calls. "I got told 'no' for two years," she says. Wondering if she should continue pushing forward, she turned, once again, to her source. "I was really frustrated," she says, "and I literally said to the universe, 'I need a sign and I need it to be really clear.'"

One April day in 1999, after wrapping up a sales training session in Detroit, Blakely went back to her hotel room and switched on *The Oprah Winfrey Show* just as Oprah was lifting her right leg to show the stockings under her gold pants. "I cut my panty hose and they're right there," Winfrey told her audience, explaining that she wanted to wear her high-heeled sandals without a seam at the toe. "I just started to cry," Sara says. "And I thought, Ok, this is the most direct sign. Thank you, universe."

Emboldened by Oprah, Blakely pushed forward with fresh resolve. She wrote her own patent (with help from a how-to book she bought at Barnes & Noble), found a hosiery factory willing to manufacture her product, and used her sales know-how to persuade buyers to meet with her. She landed her first sale after asking a buyer at Neiman Marcus to come with her to the bathroom so she could show Spanx at work on her body rather than talk about it in theory.

For the first year, Blakely avoided naysayers by keeping her business a secret from family and friends—a tactic she says warded off the inevitable question that quashes entrepreneurial dreams: If it's such a great idea, why isn't anyone else doing it? When it was time to launch,

Blakely went full monty to boost sales. At Neiman's and other stores, she relocated her product from the hosiery department to display stands at the cash register (without authorization) and asked friends around the country to buy up Spanx at their local stores (later reimbursing them with a check). To make daring ideas happen, Blakely told Guy Raz with a laugh, "you have to be someone who is willing to ask for forgiveness, not permission."

Blakely's tenacity gives emergent entrepreneurs like Jamia Ramsey, who was sitting in the audience in San Francisco, faith in their own efforts. In 2018, Ramsey launched Blendz, a dance apparel company that manufactures tights and shoes in flesh tones ranging from "Tenacious Tan" to "Confident Cocoa." The idea materialized from Ramsey's experience as a young dancer of color; she fell in love with ballet as a child, but never felt comfortable in standard pink tights and slippers. After years of dying her tights with tea bags and spray-painting her shoes, Ramsey vowed to make apparel that Black dancers could embrace. "I just want to inspire every little girl and boy out there that, hey, ballet *is* for you."

Ramsey identifies with Blakely's doggedness. When she started her company at the age of 28, Ramsey was terrified—she knew nothing about dancewear or manufacturing, and had only a few thousand dollars in savings. But she pushed ahead with all the courage she could muster to create her brand—making cold calls, setting up meetings, finding a manufacturer, testing color samples. "I think we both have that determination—I'm going to do it anyway. I'm going to be stubborn about it," Ramsey says. "I just love her energy."

Blakely was up against daunting statistics—almost half of all new businesses fail within the first five years of launch—and from a practical perspective, her enterprise seemed inconceivable. She had no M.B.A., no knowledge of manufacturing, no business plan, no budget for advertising, and just $5,000 that she had set aside from savings.

And she was only 29 years old. Although start-ups have a reputation for being the dominion of youth, a team of economists recently found that the average age for the fastest-growing new ventures is actually 45. The reason: Years of experience matter.

Jon Carroll met Blakely at a newly formed networking group of young entrepreneurs in Atlanta, where Sara had moved to build her company. The only woman out of 10 members, she was "cut from an entirely different cloth," says Carroll, who started a company that made boxer shorts with college logos out of his dorm room at the University of Virginia in 1985. During introductions, as Sara described her ambitious plans, Carroll wondered if she'd make it. "She said 'I'm running this business so it can be a platform for my overall mission to make the planet a better place for women,'" says Carroll. "We joked that we were never going to see her there next year. This is not going to survive."

But Sara had distinct attributes that helped her build her company in an unconventional way: self-awareness, deep intuition, compassion. These are traits that go back to childhood. Ford, who had a speech delay as a toddler, recalls the way his sister became his first champion. "She got me at an early age," he says. "She felt like a kindred connection, a good mentor and guide." Blakely's father recalls the way his daughter used to pick up nonverbal cues from the jury during his trials. At lunch, she'd tell him which ones to look out for. "I'd say, 'Dad, you don't have juror number four and juror number eight. They are not on your side,'" Sara says.

Blakely vividly remembers the moment she committed to embracing her "emotional intelligence," a term made popular by psychologist Daniel Goleman in the mid-1990s, as the backbone of her leadership style. It happened one night at a cocktail party when several men approached her and said they heard she'd invented something. "I said, 'Yes, I did, I'm so excited,'" she tells me, "and then one of the guys put his hand on my shoulder and said, 'Great. You know business is war. I hope you're ready to go to war, Sara.'"

Later, at home in her Hello Kitty pajamas, Blakely stared at the floor. "I thought, 'I don't want to go to war,'" she says. "And I made a decision that night that I was going to go about this business in a very different way." Rather than go into attack mode with traditional advertising, she became her own promoter, growing Spanx through word of mouth. She chose savvy over ego, following her instincts when it came to fending off hungry investors and maintaining 100 percent control over her company. And she drew on a spate of stand-up comedy she'd performed after college, allowing herself to be schmaltzy and funny.

Intent on providing a product that women wanted and needed—men had long dominated the hosiery industry—Blakely connected with her customers' insecurities and aspirations and spoke openly about her own. Business didn't have to be black-and-white and functional; it could be colorful, sexy, and indulgent. And better-fitting products made from carefully sourced fabrics could be priced higher (the original Spanx debuted at $20)—as long as they appealed to women, and especially if they had memorable style names, like Bra-llelujah, Undie-tectable Brief, and Booty Boost Leggings.

Early on, Spanx benefited enormously from celebrity testimonials—including Jessica Alba, Tyra Banks, Gwyneth Paltrow, Beyoncé, and Madonna. After Blakely shrewdly mailed a gift basket to Oprah, the talk show host chose Spanx as one of her favorite products in 2000. Six years later, with Blakely on stage as a featured guest, Oprah described Spanx as "those fabulous footless hose that suck it all in," spiking company sales of Power Panties to more than 20,000 pairs that day alone.

Carroll, who worked briefly at Spanx during a transition year in 2016, says Blakely has powerful vision, an ability to think ahead and

Spark

think big. "I've never seen anyone manage five to 10 years on the horizon better," he says. "She flies at 50,000 feet. We're all flying at 5,000." Blakely not only survived; she out-profited Carroll and the other men in the group, with whom she still meets 17 years later. Spanx does not release statistics about its annual revenue (Forbes's last estimate put it at $400 million annually), but the company has not stopped growing its products: The start-up that launched with women's shapewear also sells pants, skirts, swimwear, and men's underwear and socks.

There have been detractors along the way, some of whom have argued that Spanx is no different from our grandmothers' girdles, benefiting off a patriarchal power that requires women to look slim and shapely. But Spanx customers see her garments differently; they make them feel confident. Blakely, a feminist, says she always had the feeling that her grandmothers—even her mother—never had the chance to live their best lives, and that the core of her mission is providing opportunities for women. "I got my inspiration from what the women in my life *didn't* do and *couldn't* do," she says. "For as long as I can remember, I have had gratitude for being a woman born in the right country at the right time. I want to make the most of my opportunity."

Blakely is open about the self-doubts and insecurities that inevitably surface. "I'll be about to give a speech and the voice will be like, 'Why are you even here? What do you even have to offer these people?'" she says. She's terrified of heights and manages her fear of flying by listening to the same song (Mark Knopfler's "What It Is") and eating the same snack (Cheez-Its). At difficult moments, she seeks out close friends, plugs into Wayne Dyer, reads an inspirational quote, or embarks on a "think vacation" or a solitary getaway to recharge.

Long ago, she made the decision to be brave. "It is a choice I have made repeatedly," she says. "The only way to increase your courage is to just do the things that scare you."

Sara Blakely

When she was a child, Blakely made a pronouncement to her mother about her destiny. "Mom, I just want to let you know that when I grow up, I'm going to make all my own money," she told her, "and if I want to marry a garbage man, I'm going to."

Blakely didn't marry a garbage man. She married Jesse Itzler, a former rapper, author, and entrepreneur (he co-founded Marquis Jet, a private jet card company). The two met at a charity poker tournament in 2006 and wed two years later, when Blakely was 37 and Itzler was 40. He has described his wife as 50 percent Einstein, 50 percent Lucille Ball. During a Zoom chat, I ask Itzler what he means. She's a genius when it comes to developing, marketing, producing, and selling her products, he tells me, "but at the same time, there's a good chance she forgot her keys, locked herself in the car, and can't get out of the garage."

Blakely did, however, follow through on making her own money. An outlier among successful entrepreneurs, she continues to fend off hungry investors and still owns 100 percent of the company. Itzler, who marvels at her ability to keep Spanx unencumbered by shareholders, credits this to her keen business smarts, her ability to trust her gut, and her Clearwater upbringing. "She still has that small-town beach attitude and she operates by it. That's her guiding light. She still wears flip-flops to meetings and listens to the same music from the 1970s and '80s," he says. "I don't think she's been very affected by her success."

She has, of course, benefited from it financially. Blakely spends her money on everything from child care for the couple's four children to an annual birthday getaway with her best friends from middle school (she tells them what to wear and whether or not they'll need a passport, and surprises them with the destination).

Spark

But Blakely is saving a substantial portion of her money to give away. In 2005, after competing in the reality TV show *Rebel Billionaire*, host Richard Branson gifted Blakely $750,000 to launch a philanthropic foundation supporting women's education and entrepreneurship, including a scholarship and mentoring program for aspiring college students in her hometown of Atlanta. She committed to the Giving Pledge in 2013. During the 2020 pandemic, she gave away five million dollars in individual $5,000 grants to a thousand female entrepreneurs in the United States, and offered to loan her wedding dress to brides in need on Instagram. "Hoping this will possibly help ease someone's plans during this time," she wrote.

Blakely finds joy in friends and family and invites the rest of the world to indulge in her humanity. "She's still such a kid," Itzler tells me. "She has such a playful approach." Blakely posts photos of herself on Instagram draped in children with no makeup, her roots growing out. In one video of a morning pancake-making session, she sings Elton John's "Your Song" into a spatula microphone. "Make mistakes," she writes, "live messy, be bold." It is an authenticity that is genuine, says her brother, Ford. "Sara has not changed a bit."

That authenticity has played out for Emily Kenison. After her mentoring proposal in San Francisco, Blakely posted a photo of the two of them on Instagram, writing: "Let's give this new entrepreneur some love." Almost immediately, Kenison heard from investors and saw an uptick in interest and sales of Straplets. Every quarter, Blakely gets on the phone with Kenison, offering advice and bolstering her confidence. "She tells me to continue to bootstrap, just like she did, that I can do it," Kenison tells me. "She reminds me that it's okay not to know the answers. You just have to find them."

Blakely's old friend Jon Carroll believes she is now in the second inning of a game that has only just begun. Sara wants her legacy to stretch beyond her brand, he says. "Spanx is her platform. It does not entirely define who she is." Teaching has become a big part of her mission, and her audience is only growing. In addition to speaking at events and broadcasting lessons learned on her social media accounts, Blakely recently created a 14-part workshop on MasterClass, an online platform of video instruction taught by high-profile leaders in a variety of fields, from Helen Mirren to Neil deGrasse Tyson.

At the end of her course, Blakely summarizes her odyssey, from newbie 27-year-old entrepreneur to business leader charting the terrain for others. As I watch her tell her story, she becomes emotional and even teary about what she has been able to accomplish. I can't help but flash back to her existential ponderings—those questions she used to ask her mom at bedtime: "Why am I here? What's my purpose?"

Now five decades into her pilgrimage, Blakely seems to have figured it out. "If I have the opportunity to inspire somebody else and to encourage you to know that you have what it takes as well to go out there and give it a try, and to continue to believe in yourself," she says, "then there's nothing greater than why I'm here and what this is all about."

CHAPTER 7

Julia Child

The French Chef (1912–2004)

At the beginning of her cooking life in Paris, when Julia Child first fell in love with the smell of shallots sautéing in fresh butter and the luxurious taste of a fresh baguette, she had a bit of a hardware problem. Although the two-floor apartment she and her husband, Paul, had rented at 81 rue de l'Université in the 7th arrondissement came equipped with a functional stovetop oven, the apparatus was notably short.

In a famous black-and-white photo Paul took in 1953, a six-foot-two Julia—apron tied around her waist, left hand on her hip—towers over the pot she is stirring. With the diminutive stovetop barely reaching the middle of her thigh, Child looks like a grown-up amusing herself on a toddler's kitchen play set.

This mismatch in height plagued Child throughout the 13 years she lived in France. "If we ever get into the money," she wrote to a

Spark

friend in her typically blunt (and not always politically correct) manner, "I am going to have a kitchen where everything is my height, and
none of this pygmy stuff."

In 1961, when the Childs moved into their home in Cambridge,
Massachusetts, Julia finally got her kitchen, "the soul of our house," as
she would later describe it. Paul saw to the Julia-perfect design. Pots
and pans hung from pegboards so they'd be quickly accessible and easy
to reach. Knives were positioned near the sink, and utensils were close
to the stove. Spotlights hung from the ceiling to illuminate the workspace. And most important of all, maple countertops were installed at
an unusual 38 inches—two inches higher than standard.

Julia Child's distinctive kitchen is now a tribute to her legacy. In
2001, she donated the entire room to the Smithsonian National
Museum of American History, where it has been reassembled in its
original form with every accoutrement: the painting of an artichoke
above the oven; the cheese knives and cherry pitters; the stone crab
claw cracker; the crepe pans and fish mousse mold; the ceramic bowl
from southern France; and even the banana stickers Paul used to stick
to the underside of the dining table in the morning after peeling his
fruit for breakfast.

The day I visit the exhibit, I find Julia fans oohing over the retro
appliances, marveling at Julia's height in a life-size cardboard cutout,
and lovingly mimicking her distinctive high-pitched voice. A docent
standing in front of the kitchen window explains to his tour group
that even the clock above Julia's sink is stopped precisely when curators
began packing the room for shipment to the museum for public
viewing: 12:19 p.m.

"Look, there she is!" one visitor says excitedly as she moves toward
a TV monitor playing video clips of the "French chef" in all her gas-
tronomic splendor—whacking a piece of fish here, taking a blowtorch
to a tomato there. In one scene, she pats the derriere of a suckling pig

before roasting it; in another, she's wearing a yellow rain slicker and holding a bright green umbrella to ward off sprays of water as she washes lettuce. When she trills, "I'm Julia Child," I think, "Yes, she is. No one else can do *that!*"

Julia Child was beloved for a host of reasons: her down-to-earth demeanor, her warbling intonations, her ability to exhaustively explicate cheese soufflés and beef bourguignonne with a dash of raunchy humor. But one standout attribute made Child irresistible above all the others: her middle-agedness.

A culinary neophyte well into adulthood, Child did not discover food until she famously consumed her first meal at restaurant La Couronne in Rouen, France, at the age of 36. It was an awakening that sparked a midlife passion and defined her purpose in life. She enrolled in Le Cordon Bleu in her late 30s and co-wrote her best-selling cookbook, *Mastering the Art of French Cooking*, in her 40s. At the age of 50, when many of her peers were hunkering down for retirement, she debuted as a television cook on her pioneering show, *The French Chef.*

Neither child prodigy nor late bloomer, Julia Child ranks as what we might call a midlifer—a person who discovers her destiny as the halftime buzzer sounds. There were times when Child bemoaned her late entrée into the career she loved. "I wish I had started in when I was 14 years old," she lamented when she was 39. But Child found her passion at a time when *she* could flourish. Midlife allowed her to court her fans with an openness, wackiness, and relatability that kept them coming back. It was midlife that launched what would become her extraordinary 50-year career.

When writer Alex Prud'homme, Child's grandnephew and biographer, asked if she knew what an enormous impact she'd had on cooking in America, "Julia shrugged and demurred," he recounted in his book, *The French Chef in America.* "'Well, if it wasn't me, it would have been someone else,'" she said. "But it *was* her," Prud'homme continued, "And it is unlikely that anyone else could have done what she did, when she did, and how she did it. Julia Child changed the nation, even if she didn't like to admit it."

⌒⌒

The year 1912 spawned a flock of American luminaries: Dancer Gene Kelly, civil rights leader Dorothy Height, folk musician Woody Guthrie, photographer Gordon Parks, humorist Minnie Pearl, and celebrity French chef Julia Carolyn McWilliams, born in Pasadena, California, on August 15—just a few months before the Boston Red Sox beat the New York Giants to clinch the World Series, and Democrat Woodrow Wilson ousted incumbent Republican President William Taft in a landslide.

Julia's mother, a spirited redhead named Julia Carolyn Weston, was one of 10 children who came from a family of old money in western Massachusetts. Caro, as she was called, met John McWilliams, Jr., a tall Princeton grad from a banking family, through a friend in 1903. After a lengthy eight-year courtship (drawn out by Caro, who spent much of her time trekking to health spas with a sister who had tuberculosis), the couple married in January 1911 at St. Stephen's Episcopal Church in Colorado Springs. From there, they moved 1,000 miles west to settle in Pasadena, California.

The McWilliams family, which soon included Julia's younger brother, John, and sister, Dorothy, lived a comfortable life in a house that came with a sleeping porch, a tennis court, and an abundance of orange and avocado trees. A rambunctious kid, Julia and Babe Hall, her prankster neighborhood sidekick, kept busy when they weren't in school, tearing through the streets on their bicycles, throwing mud pies at cars from the roof of a neighbor's garage, and stealing cigars from their fathers. Julia, a friend later noted, was often the instigator.

Some people must choose first jobs, if not careers, out of financial necessity. Julia was fortunate not to feel that pressure. Her college path was mapped long before she graduated from the elite all-girls Katharine Branson School in Ross, California. Caro was a proud Smith College graduate, class of 1900, and there was never a question that her firstborn daughter would be, too. "I was entered the day I was born," Julia later quipped in an interview with the *Smith Alumnae Quarterly,* "and couldn't have avoided it I'd wanted to."

At Smith, Julia spent much of her time simply growing up. She majored in history, but fantasized about becoming a famous novelist. "I was a pure romantic, and only operating with half my burners turned on," she later recalled. It wasn't solely lack of ambition. In the 1930s, women were often pigeonholed into gender-designated careers like nursing or secretarial positions, rather than being encouraged to think big. And there was always an underlying expectation that marriage and motherhood would take priority.

"Like most women of my era," she once said, "I was truly prepared for nothing."

From childhood, Julia Child had a healthy love affair with food. It didn't matter that the repertoire at home was fairly limited. Her father didn't cook and her mother's meals consisted of a standard menu: baking powder biscuits, Welsh rarebit, and cod fish balls with egg sauce. Much of the rest of the time, a hired cook prepared dinner, which included a hearty dosing of meat and potatoes. Julia gobbled it down.

At Smith, she indulged in jelly doughnuts and, toward the end of Prohibition, discovered gin and tonics at nearby speakeasies. "I don't remember anyone dieting or weight-watching or having eating disorders," she said. "There were no vegetarians and no nutrition police. We just ate. Everything. The more the better."

And she could afford to indulge. By the age of nine, Julia towered over her classmates; by adulthood, she and both of her siblings would soar past six feet, prompting their mother to joke that she "raised 18 feet of children."

Food was pleasurable, but never a professional aspiration. "My interest in cuisine did not extend beyond my huge appetite," Julia said. Instead, after graduating from Smith in 1934, she applied for editorial positions at the *New Yorker* and *Newsweek,* hoping that learning about writing might help her craft her career. But neither magazine offered her a position. Her first job, at W. & J. Sloane's furniture store on Fifth Avenue, paid her $18 a week to write press releases and catalog copy.

Career paths rarely lead neatly from A to B. Often, as in Julia's case, there's a directionless period—a time of uncertain rambling when insecurities surge and doubts begin to plague. Life's misfortunes intervene, too. Julia loved living in New York City, but the job itself was not fulfilling, her first serious boyfriend broke up with her, and her mother's health began to fail from severe complications of high blood pressure. Home beckoned, and a disheartened Julia decided to return. "It was one time I can remember feeling hopelessly lost," she told biographer Bob Spitz.

In Pasadena, Julia helped look after her mother and was at her bedside when Caro died at the age of 60 in July 1937. Her mother's death was crushing, and over the five years that followed, Julia existed without purpose. She and her conservative father whose political views clashed with her own often spent time on the golf course—one of the few places the two of them got along. She attended her sister's graduation from Bennington College and did some volunteer work with the Red Cross and the Junior League.

As the months ticked by, Julia's post-college hiatus became mired in uncertainty. What was she meant to do? "I knew I didn't want to become a standard housewife, or a corporate woman, but I wasn't sure what I *did* want to be," she wrote in her memoir. She didn't strive all that hard professionally—at least not by any traditional standards. At one point, she wrote a monthly column about Southern California fashion for a start-up San Francisco magazine, but the publication went bankrupt after a year. And a stint with W. & J. Sloane's West Coast office ended after Julia was fired for failing to make minor changes in ad copy text for a furniture sale.

Julia had a nagging feeling that she should be doing more. But she was uneasy about her abilities and uncertain about whether she had the goods to succeed. In 1940, just before her 28th birthday, she confided this unsettled feeling in her diary: "When I was in school and later, I felt I had particular and unique spiritual gifts. That I was meant for something, and was like no one else. It hadn't come out yet, but it was there, warm and latent," she wrote. "Today, it has gone out and I am sadly an ordinary person."

⌒

Passion seems to inhabit child prodigies. They show early talent and interest in art or music or math, and keep at it because they must. It lurks in their every cell. It *becomes* them.

But what about the rest of us? What was Julia's purpose in life? And why hadn't she found it?

Self-discovery requires peeling back the social and emotional layers that shroud our souls: expectations, pressures, insecurities, fears. In the simplest way, this reminds me of one of my beloved childhood traditions: a British game called "pass the parcel." At birthday parties, we would sit cross-legged in a circle and hand a gift, wrapped in multiple layers of paper in different colors and patterns, from one friend to the next while music played. When the music stopped, the person holding the present would unwrap a single layer and pass it on. The game continued in musical fits and starts until the last layer was removed, and the gift inside was finally revealed.

Identifying one's passion is never easy, because in real life these layers can be difficult to identify and even harder to cast off. They stick like varnish on old furniture. Édouard Manet, the French painter born in 1832, was expected to follow in the footsteps of his father, an official at the Ministry of Justice in Paris. Manet applied to the Naval College when he was 16 years old, but failed the entrance exam and couldn't live up to the plan. He loved art, and when it became clear that this desire was what drove him, his father relented. Manet went on to become a pioneering artist of the modernist era.

These kinds of stories assume, of course, that our purpose in life is embedded within us and needs rooting out. But what if it is something we *develop*, instead? Carol Dweck, a professor of psychology at Stanford University, has written extensively about the difference between what she calls a "fixed mindset," which assumes that intelligence and potential are cast in stone, and a "growth mindset," which allows for cultivating abilities and interests over time. Dweck's research has made inroads at schools, where students are encouraged to embrace the idea that mistakes can provide valuable lessons and positive thinking: "I can be good at this if I keep trying," rather than "I am terrible at math."

When it comes to passion, the growth mindset suggests that finding one's calling requires curiosity, openness, and a willingness to explore. Dweck and a team of colleagues tested the "find your passion" mantra among college students. They found that those who have fixed ideas about their abilities and what they're meant to do give up more easily than people who are more open to change. In one of their experiments, the researchers tested "techie" students (interested in science, technology, and mathematics) and "fuzzy" students (arts and humanities) and found that those who held fixed mindsets were less open to absorbing information from articles on subjects outside of their interest area.

The problem, the researchers suggest, is that an unwillingness to explore new terrain can limit potential and result in disappointment in the long run. "Urging people to find their passion may lead them to put all their eggs in one basket but then to drop that basket when it becomes difficult to carry," they concluded in their 2018 study.

In Julia Child's case, passion would be sparked by a sensuous experience: the smell and taste of a rapturous meal in northern France. But as a late 20-something boomerang kid, drifting back home in Pasadena, she had no idea what was coming or when. She needed to do something—something that would engage her ebullient and vigorous self, bolster her confidence, and lay a footpath to the future.

The war provided the opportunity. In December 1942, she landed a job at the Office of Strategic Services (OSS) in Washington, D.C., a newly formed government intelligence agency. The position itself wasn't all that stimulating—she spent much of her time organizing and filing paperwork—but Julia reveled in the camaraderie of bright, motivated co-workers. And working for a clandestine operation (OSS's nickname, at the time, was "Oh So Secret") added an element of excitement.

OSS's mission and Julia's colleagues made her want to stay, and she made the most of the opportunity. Responsible and motivated, she discovered that she had an organized mind—an ability that would serve her well as a recipe tester years later. When the opportunity arose to go overseas with the agency, Julia raised her hand and was transferred to an OSS office in South Asia. She had a yearning for adventure that Nancy Verde Barr, a chef who worked with her for two decades, suspects Julia inherited from her pioneering paternal grandfather, who was a teenager when he left Pike County, Illinois, in 1849 to join the gold rush in the Sacramento Valley.

In March 1944, Julia boarded the S.S. *Mariposa*—the same luxury liner that had taken seven-year-old Shirley Temple to Hawaii in 1935 and had since been turned over for war duty. Along the way, she wrote for the ship's daily newspaper, and kept up with the goings-on of the nine women and more than 3,000 men aboard. After 31 days on water, Julia landed in the former British colony of Ceylon, now Sri Lanka, and made her way to the mountainous town of Kandy.

The OSS office, housed in a colonial estate on a tea plantation, was nestled next to a lake and surrounded by banyan trees, rice paddies, and Buddhist temples—a far cry from Pasadena, let alone Washington, D.C. Having earned a high security level clearance, she was now the organizer of classified documents filed by field agents in the region. She soaked up the intellectual banter with a whole new cadre of colleagues, played tennis, and checked out the nightlife in nearby Colombo.

There are times in life when a single chance encounter determines a life journey. For Julia, that moment came in May 1944, when she met Paul Child, an OSS officer who had been sent to Kandy to create war maps. Initially, Julia found Paul pleasant, though not especially good-looking. At 42, he was 10 years older, cerebral, fastidious, and far more worldly. In letters home to his twin brother, Charlie, Paul was reticent as well. Julia was "warm and witty," he told Charlie, but somewhat high-strung. "Her slight atmosphere of hysteria gets on my nerves," he wrote.

As they got to know each other, however, their dissimilarities became points of intrigue. Within a year of meeting in Ceylon, Paul was transferred to Chongqing, a city situated at the confluence of the Yangtze and Jialing Rivers, and Julia to Kunming, the capital of China's

Julia Child

Yunnan Province, where subsequently Paul was reassigned. China brought them together over sizzling woks and chopsticks. Paul, who had lived in Paris for five years, was already a gourmand; Julia, an adventurous eater, relished the exotic flavors of duck feet and clay pot spicy chicken stews. "Food," writes biographer Spitz, "seemed to be their common aphrodisiac."

Fun soon developed into devotion. For her 33rd birthday in August 1945, Paul penned Julia a love sonnet. Weeks later, with the war over, they shipped back to the United States and soon took a cross-country escapade in Julia's Buick from Pasadena to the rugged coast of Maine to visit Charlie and his family. After a few days, Julia reflected, "We took deep breaths and announced: 'We've decided to get married.'"

Paul's family had an enthusiastic response: "About time!"

～～

Julia Child was 36 years old when she and Paul boarded the S.S. *America* bound for Le Havre harbor on the Normandy coast. France began working its magic immediately: The bicyclists, the children in wooden shoes, the pop of green from cabbage fields. By the time the Childs arrived in the town of Rouen—about 60 miles northwest of Paris, where Paul was starting a new job with the U.S. Information Agency—Julia was smitten.

In this ancient town Julia would, finally, discover her raison d'etre. Neither an intellectual revelation (like Newton's understanding of gravity) nor a decision of intent (like Sara Blakely's pursuit of an entre-preneurial idea), it was instead an awakening borne of inquisitiveness, openness, awareness, and luck. Eating food had appealed to Julia since childhood, and unfamiliar tastes in China had roused her palate. Now, in a country where citizens honored their cuisine as fervently as their national flag, Julia's taste buds were primed for discovery.

Spark

Trusting their *Guide Michelin* as culinary adviser, Julia and Paul chose Rouen's celebrated La Couronne for lunch. Julia drank in the pungent aromas of buttery shallots and fresh lemon mixed with wine vinegar before feasting on a first course of briny oysters served with rye bread and butter. Then came the main entrée: a Dover sole, known as *sole meunière*, served whole on an oval platter.

"I closed my eyes and inhaled the rising perfume. Then I lifted a forkful of fish to my mouth, took a bite, and chewed slowly," Julia recollected. "The flesh of the sole was delicate, with a light but distinct taste of the ocean that blended marvelously with the browned butter. I chewed slowly and swallowed. It was a morsel of perfection."

A *salade verte*, fresh crispy baguette, and *fromage blanc* topped off the meal, which Julia would later describe as an epiphany—the moment when she knew that she was meant to be a cook.

It was a moment she never forgot. "In all the years since that succulent meal," she wrote, "I have yet to lose the feelings of wonder and excitement that it inspired in me. I can still almost taste it."

∽

France has a reputation for electrifying the senses. In *A Moveable Feast*, his memoir about living in Paris in the 1920s, Ernest Hemingway recounts the taste of *pommes à l'huile* (potatoes marinated in olive oil) and *tournedos*, small filets of beef dripping in sauce *béarnaise*. Restaurateur Alice Waters described her first experience with French food as a revelation. "I discovered fresh baguettes, apricot jam, oysters right out of their shell, wild rocket salad from Nice, crêpes from Brittany and felt like I had never eaten before," she once said.

For Julia, shopping in the Paris markets introduced her to the lure of freshly killed rabbits and the thrill of a proper *turbotière* in which to cook fish. She learned the importance of understanding the proper pairings of meats and fish and vegetables. Knowing what kind of cheese you wanted to buy from your neighborhood *crémerie* wasn't enough— you had to know at which meal you were planning to serve it. Food, she discovered, was as much about relationships as about eating.

For her 37th birthday, in August 1949, Paul bought her the *Larousse Gastronomique,* France's 1,087-page bible of cooking, which Julia devoured. She began experimenting with recipes for everything from turnips to escargot in the couple's rue de l'Université apartment (affectionately nicknamed "Roo de Loo"). Some friends questioned her newfound obsession, but Paul stood by steadfastly, encouraging her to keep moving forward. "By now, I knew that French food was *it* for me," she recalled.

Through her 20s and early 30s, Julia had enjoyed a privileged life, subsisting without a clear plan for the future. Now that food had taken hold, she became purposeful about her training and enrolled at Le Cordon Bleu culinary institute in the fall of 1949. A rotund Chef Max Bugnard taught his students how to prepare crudités, scrambled eggs, roasted partridges, and beef bourguignonne. Bugnard not only explained ingredients, but also reinforced the traditions, focus, and creativity required to prepare a recipe. "How magnificent," Julia later declared, "to find my life's calling, at long last!"

In letters home to his brother, Paul also delighted in his wife's transformation. "All sorts of *délices* are spouting out of [Julia's] finger ends like sparks out of a pinwheel," he reported. At times, she seemed to be a musical conductor—stirring pots in tandem as she rhythmically opened and shut the oven. "Warning bells are sounding off like signals from the podium," he wrote, "and a garlic-flavored steam fills the air with an odoriferous leitmotif."

In 1951, three years after moving to Paris, Julia met Simone Beck Fischbacher, a fellow Cordon Bleu graduate, at a party; the two discovered their mutual passion for cooking. Through Simca, as she was known, Julia was introduced to foodie Madame Louisette Bertholle. Simca, raised in Normandy, was more than five feet eight inches tall, exuberant, and a dessert lover; Louisette was shorter and quieter—"a dear person, small and neat, with a wonderfully vague temperament," as Julia described her.

Bonded over their love of cuisine, the trio decided to open a home cooking school at Roo de Loo, which they would call L'École des Trois Gourmandes. On January 23, 1952—three years after Child's meal at La Couronne and less than a year after her graduation from the Cordon Bleu—the women convened their first class. In what would turn out to be a prelude to her later TV cooking shows, Julia mined her newly acquired culinary expertise as she and her collaborators shared their skills and recipes with others. At 10 a.m. on Tuesdays and Wednesdays, students arrived and made watercress soup—or whatever else was on the menu that day—for a sumptuous sit-down lunch.

Through teaching, Julia discovered that published recipes were often imperfect; she began writing down tweaks on paper, building a collection of recipes of her own. Simca and Louisette, meanwhile, approached their friend with a request. The pair had drafted a French cookbook, which they hoped to publish in the United States, but their American editor had quit and they needed a replacement. Might Julia help get their manuscript, rife with bungled English and a mishmash of recipes, to the finish line?

The opportunity landed with a thud. Simca and Louisette's 600-page manuscript needed a complete overhaul, not just an edit. But Julia took on the challenge with gusto, rearranging text, clarifying convoluted instructions, and rewriting awkward language. The bigger job was fixing the bones of the thing. She attacked every one of the recipes like a scientist, double-checking each egg yolk and stalk of celery, and subjecting every instruction to what she called "operational proof."

Recipe testing takes time, especially if you are exacting about your work. Julia's meticulous edit took seven years and followed her from Paris to Paul's subsequent postings in Marseille, West Germany, Washington, D.C., and Norway. Along the way, Louisette bowed out of much of the substance of the work, allowing Simca and Julia to develop into a creative duo. The two delighted in their collaboration while also weathering spirited disagreements—often when Simca's devotion to French tradition clashed with Julia's desire to reinvent it for Americans. By the time they agreed to the last pinch of salt, the book was an exorbitant 850 pages.

Houghton Mifflin, which had expressed interest in the book, rejected it twice—first at its original length, and again after it was reduced to a still portly 684 pages. In the summer of 1959, the manuscript made its way to the desk of Judith Jones, a young food editor at Knopf, who saw promise in the detailed instructions for everything from mixing egg whites to seeding a tomato.

Jones liked the structure of the book, with its master recipes and variations that followed, as well as the overarching message: French cooking should be accessible and fun. "From the moment I started turning the pages, I was *bouleversée,* as the French say—knocked out," Jones later recalled in her memoir, *The Tenth Muse.* "This was the book I'd been searching for." It was, she said, a "work of genius."

In October 1961, *Mastering the Art of French Cooking* debuted with a rave from *New York Times* critic Craig Claiborne. Although he had a few nitpicks—the book had no recipes for puff pastry or croissants (those would come in a later volume)—he described the recipes as "glorious" and the work "laudable and monumental."

As *Mastering* went into a second printing, Julia and Simca set out on a six-week book tour and made the publicity rounds of TV interviews. In New York City, James Beard, America's culinary trendsetter at the time, hosted a celebratory dinner for the authors at the

Spark

Egg Basket restaurant in Manhattan. "The high point of the evening," Julia later reflected, "came when Jim Beard stood up and toasted me and Simca with the highest compliment imaginable: 'I love your book—I only wish that I had written it myself!'"

In early 1962, Julia received an invitation that would jump-start her celebrity career. *I've Been Reading,* a public television show about books hosted by a college English professor on Boston's WGBH, invited her to be a guest on their hour-long show. Realizing she needed to fill airtime, Julia brought with her a hot plate, eggs, mushrooms, an apron, and a lesson in how to make an omelet. At the time, she was known only as a name on the cover of a book. But Julia's entertaining TV cooking delighted viewers who had never seen anyone like her—or anything like her enthusiasm for food. Immediately after airing, more than two dozen fans wrote to the station with a request: more of that "tall, loud woman cooking on television."

One year later, after Paul had retired from government work and he and Julia left Europe to settle down in Cambridge, Massachusetts, WGBH eagerly offered Julia her own weekly cooking show, *The French Chef.* In November 1966, she vaulted to the cover of *Time* magazine. By then, her impact had reached cult status.

"Manhattan matrons refuse to dine out the night she is on," *Time* reported. And even the president couldn't compete. "When Washington, D.C.'s WETA interrupted her program to carry Lyndon Johnson live, the station's switchboard was jammed for an hour."

Julia was 49 when *Mastering the Art of French Cooking* debuted; 50 when she launched her TV show; and 54 when *Time* hailed her as a culinary hero. She was deep into middle age, but only beginning a professional odyssey that would grow ever bigger over the next four decades.

Julia Child

Midlife has long been considered a period of angst. Youth is gone, death is nearing, and life morphs into a series of question marks: How do I put aging on hold? Is my career almost over? Is *this* all there is? Often, the day-to-day seems hollow, even futile.

In 1965, Canadian psychoanalyst Elliott Jaques coined the infamous term "midlife crisis" in a report in which he described an age at which things start to sour, around a person's mid-30s. Jaques based his findings on a study of 310 highly accomplished male composers, artists, and writers, many of whom experienced declines in their creative output around the age of 35. Rossini, for example, wrote his first opera, *The Marriage Contract,* in his teens, and completed two dozen more—including his famed *The Barber of Seville*—by the time he was 31. He composed his final operatic work, *William Tell,* the year he turned 37.

Even more striking, Jaques noted, were the notable number of deaths that occurred around midlife. The painter Raphael, the poet Arthur Rimbaud, musicians Chopin, Mozart, and Henry Purcell—all died in their mid- to late 30s. "The closer one keeps to genius in the sample," Jaques wrote, "the more striking and clear-cut is this spiking of the death rate in midlife." Midlife, it seemed, was not only a period of decline, but also the moment of ultimate demise.

In 1976, journalist Gail Sheehy catapulted Jaques's academic discourse into a cultural phenomenon with the publication of her book, *Passages: Predictable Crises of Adult Life.* Everyone, women included, seemed to experience some kind of lull in midlife. Moods dipped, behaviors changed, affairs were had, and sports cars purchased. Creative juices and career ambitions, meanwhile, were pushed into a stagnating stew of regrets about the past and worries about the future.

New research has since challenged the notion that happiness dips at a prescribed halfway point in life. But Julia's blossoming occurred during its heyday in the late 20th century—and she was clearly an outlier at the time. She seized professional opportunities with the same gusto she used to plunge lobsters headfirst into boiling water. And she kept climbing.

Alex Prud'homme described his great-aunt's middle years as a transformative time, when she discovered "cookery," as she liked to call it, and figured out how to excel at it. "This was her gestational period, when she was in her thirties, forties, and early fifties," he wrote. "I think of it as Act One of Julia's adult life."

Julia's second act, as Prud'homme sees it, stretched from the late 1960s to the early 1980s. During these years, Julia's enterprise exploded. In 1970, she and Simca produced the second volume of their best seller; more books followed. *The French Chef,* which would eventually amass some 200 episodes, evolved into color TV, and Julia became a regular guest on *Good Morning America.* At age 71, long past conventional retirement, Julia launched *Dinner at Julia's,* a television series featuring guest chef celebrities, including Wolfgang Puck and James Beard.

Timing is always an ingredient of success. During Julia's childhood, cooking in America was limited by technology. Electric refrigerators did not become available to consumers until the 1920s, making storage of fresh ingredients challenging. And certain foods hadn't even been discovered; the ruby red grapefruit, for example, was a chance mutation farmers stumbled upon in 1929. If anything, the culinary trend launched during Julia's early years was moving toward simplicity in the form of processed foods, like mayonnaise and hot dogs. Prohibition's ban on alcohol between 1920 and 1933, meanwhile, put a damper on the pleasure of combining food and drink.

Julia's chance meal in Rouen came at an opportune moment. Americans were beginning to travel abroad in greater numbers after World War II, opening their minds and whetting their appetites for new culinary adventures. *Gourmet* magazine published its first issue in 1941, and the Culinary Institute of America opened doors at its original location in New Haven, Connecticut, in 1946. By the time *Mastering the Art* debuted, enrollment was burgeoning, and the Kennedys, newly ensconced in the White House, had hired a French chef. "America was ripe for a new culinary age," Barr writes in her memoir, *Backstage with Julia*.

The media also helped bolster Julia's rise in myriad ways. Scientists become known as they publish papers and politicians as they give speeches. Visual technology was a game changer for Picasso, who used photography to bolster his image. Shirley Temple's stardom was tied to the evolution of Hollywood film. Twenty-first-century personalities like Sara Blakely use social media to tell their stories; her adept Instagram posts have drawn tens of thousands of followers and built brand loyalty.

For Julia, television gave a little-known cookbook author the platform to become a superstar. And she used the medium to her advantage. With a unique blend of unpretentiousness (she was a "cook," not a "chef") and schmaltz, she made cooking a bungling adventure, not a chore. She patted her ribs when talking about chicken parts, waved her rolling pin around, and gleefully admitted mistakes. After flubbing a potato pancake flip—it fell apart when it landed in the pan—she said simply: "Well, that didn't go very well. You see, when I flipped it, I didn't have the courage to do it the way I should have."

Midlife, with all its imperfections, became Julia's secret weapon. She looked like a fairly ordinary woman, only taller, without glamour or glitz. Her hallmark was her ebullient "I-can-do and you-can-too" attitude, which enveloped her viewers with the confidence that they, too, could whip up a lamb, goose, and sausage cassoulet.

"Having discovered cooking 'late,' while in her thirties," Prud'homme writes, "Julia empathized with her viewers and was unafraid to ask obvious, 'dumb' questions, which made cooking

Spark

comprehensible: What's the best way to boil an egg? How do you make a chocolate cake? What kind of wine should I serve with cheese?"

Julia's genius was in knowing instinctively that although proper instruction was critical, it also had to be entertaining. She flirted with male guests and peppered her dialogue with off-color jokes. Stephanie Hersh, Julia's longtime assistant, remembered one such moment at a cooking demonstration at the Long Wharf Theatre in New Haven. In response to an audience member's question about why she wasn't using extra virgin olive oil to sauté her chicken, Julia replied that olive oil should be used in its raw form in a salad dressing or as an embellishment. And then, her eyes sparkling, she quipped, "Everyone knows that a heated virgin just doesn't work very well!'"

Blunt, opinionated, and stubborn, Julia refused to be co-opted by anyone, including big-dollar television sponsors. Her friend, Chef Jacques Pépin, remembered her deliberately snubbing Kendall-Jackson Winery on a show they helped sponsor. "On air, I asked Julia if she wanted a merlot, my preference, or a cabernet sauvignon with her lamb," Pépin recalled. "Looking directly into the camera, she replied, 'Beer.'" Before he knew it, she'd pulled a bottle of Samuel Adams out of the fridge. When it was Land O' Lake's turn, Julia used Crisco along with butter to make her chicken potpie dough and apple galette.

Work drove her. In addition to the hundreds of television episodes she hosted, Julia wrote or co-wrote more than a dozen books—all beginning the year she turned 50. She was tireless, often working from 5:30 a.m. to 10:30 p.m., and then declaring: "Time for dinner!" "When you rest, you rust," she once said, and "When you stop, you drop."

In her memoir, Barr recounts the frenetic pace of her first day working with Child in New York City. It started in the early morning with a TV appearance on *Good Morning America* and ended with autograph signings after a cooking demonstration (paella and *sabayon*, a light Italian custard) for about 300 guests at the famed De Gustibus

Cooking School late in the evening. Despite an early wake-up call the next day, Julia insisted on a meal—so off they traipsed to Mercurio's, an Italian restaurant, for an expansive meal with wine and amaretti cookies. Julia topped it all off with a fully caffeinated cup of coffee. Once, when a woman sympathetically said she must be tired, Julia blurted out: "I *don't* exhaust."

In her 70s, Julia allowed a few naps, but timed them to eight minutes. At 77, she took a trip to San Francisco, two days after the 1989 earthquake, to promote her latest book, *The Way to Cook*. She launched a gastronomy degree at Boston University with her friend Jacques Pépin the year she turned 79. And in celebration of her 80th birthday, Julia hopped from one celebratory party to the next over the course of her birthday year—some 300 in total.

"For Julia," Barr writes, "there was no waning culinary passion, no disillusionment, and definitely no retiring."

One fall evening in 2018, I made my way to Schola, a cooking school in the Mount Vernon neighborhood of Baltimore, for a sold-out "Tribute to Julia Child." More than two dozen people had come to learn how to make a classic Julia menu: potato leek soup, beef bourguignonne, parsley potatoes, roasted asparagus, and an apple clafoutis.

Standing in front of white ceramic bowls filled with fresh oranges and eggs, Chef Jerry Pellegrino welcomed his guests, filled our glasses with cabernet sauvignon, and introduced the celebrity chef of the evening. "Here's the goofiest woman you've ever seen with the funniest voice in the world throwing food around on television and telling everyone, 'you can do this in your home,'" he said. "No one had ever seen someone like her before."

Spark

Pellegrino had once had the pleasure of hosting Julia Child at his former restaurant, Corks. Eager to spice up her evening, he and his staff created a crown out of wine corks, which they intended to give her before dinner. Her advance person told Pellegrino not to be disappointed if Julia declined to wear it—after all, he was told, she was getting her hair done that afternoon.

Julia was not deterred. "Within seconds," Pellegrino said, "it's on her head and she's hugging me and asking to take pictures." A photo shows her beaming with the crown firmly nestled in place. "She was one of the most gracious guests I've ever entertained."

With Edith Piaf, one of Julia's favorites, playing in the background, we washed our hands and donned our aprons. For the next hour, we chopped onions and apples, cut mushrooms, simmered beef, pureed soup, and fried leeks for garnish. We also talked and entertained each other with stories. "What you'll see tonight is 26 people who don't know each other become best of friends," says Pellegrino.

Julia would have been pleased. Friends and family were the sustenance that cooking provided. She and Paul, who never had children ("they had tried, but it 'didn't take,'" she told Prud'homme), had a bevy of close friends who sat around their table for meals and conversation. And their own relationship was formidable—they were a team, often signing their letters "JP" or "Pulia." Throughout her career, Paul was her steadfast supporter, cheerleader, and photographer. They shopped, laughed, cooked, drank, and got feisty together. One year, while living in Germany, they sent a Valentine's Day card featuring the two of them sitting in a bubble bath with the headline, "Wish You Were Here."

At midlife, Julia had the freedom to dedicate all her time to her career. There were no demands from children or aging parents (her mother died before Julia joined the OSS and her father, less than a year after *Mastering,* Volume 1 was published). Even after Paul became ill—he sustained several small strokes along with other health challenges—and settled into an assisted living facility, Julia continued to work. After his death in 1994 at the age of 92, work became her refuge.

Like Shirley Temple, Julia Child had an extraordinary capacity to weather life's stumbles, and the will to keep on going. When she was 90, she told the *Los Angeles Times,* "You don't have to retire nowadays, do you? I don't even know what it would mean."

Julia had planned to mark her 92nd birthday with friends and family on August 15, 2004. A few days prior, her assistant served her onion soup and she got into bed with her beloved cat. She died in her sleep of complications from kidney failure on August 13, just two days before the festivities.

The guests who had come to join her celebrated instead "with tears and laughter and songs and stories, and plenty of good food and wine," writes Prud'homme. "She had done it again, attendees agreed: Julia had managed to exit her life as she had lived it, with a touch of drama and exquisite timing."

And with a legacy that continues to make an impact. As we ate our last bites of the sumptuous apple clafoutis we prepared in our Schola cooking class, a fellow participant took a sip of wine and expressed her gratitude to the grande dame we had come to celebrate. "She broadened our palate in so many ways," she said. "She was revolutionary."

CHAPTER 8

Maya Angelou

The Memoirist (1928–2014)

pril 4, 1968, began with celebratory anticipation for Maya
Angelou. To celebrate her 40th birthday with friends, the
writer and civil rights activist had prepared a festive meal:
Texas chili, baked ham, candied yams, macaroni and cheese, rice and
peas, and a pineapple upside-down cake. The ice buckets were stocked;
the apartment danced with daffodils. "I was really putting on the dog,"
she later recalled.

But not long before her guests were due to arrive, Angelou's friend
Dolly McPherson knocked on her door with the news that shattered
the nation: "Martin Luther King was shot," McPherson told her.
"Maya, he's dead."

Angelou, who had signed on to work with King on his upcoming
Poor People's Campaign—an effort he was organizing to demand
economic security for people living in poverty—withdrew in solitude

and grief. Already, she was shattered by the deaths of John F. Kennedy and Malcolm X, and the haunting images of their widows. Now it was MLK. "Everywhere I turned, life was repeating itself," Angelou wrote. "Depression wound itself around me so securely I could barely walk, and didn't want to talk."

It would take her dear friend James Baldwin—the poet and writer she called Jimmy—to lift her out of her despair. He came to her apartment, banged on the door, and insisted she come out with him. He was taking her to dinner at the home of his friends, Jules Feiffer, the cartoonist, and Jules's then wife, Judy, a photographer, novelist, and book editor. "They are both funny," Baldwin told her, "and you need to laugh."

That spring evening in the Feiffers' dining room, Angelou began to unwind. She regaled the group with anecdotes about the rocky landscape of her life—starting with her impoverished upbringing in Stamps, Arkansas. Her raw and engaging storytelling piqued Judy's interest; the next day, she called Random House editor Robert Loomis to tell him about the raconteur she'd met. In an interview for the 2017 PBS documentary *And Still I Rise,* Loomis recalled Judy saying about Angelou: "She's got a book in her of some kind."

Angelou had been writing since her early 30s, when an emotional breakdown prompted her singing coach to hand her a pencil and legal pad, bidding her to put her feelings on paper. She began with verses and dialogue but had yet to attempt a narrative. When Loomis called to ask if she'd consider writing an autobiography, Angelou turned him down. "I am a poet and playwright," she said. "Maybe in 10 or 20 years."

But Loomis refused to give up. In a follow-up call, he used a backhanded tactic to win over Angelou. "[Y]ou may be right," he told her, "because it is nearly impossible to write autobiography as literature." The dare galvanized Angelou, and she agreed to give it a go. "When do you think you can start?" Loomis asked. "I'll start tomorrow," she said.

Now 40 years old—just shy of what would turn out to be the midway point in her life—Maya Angelou began to unravel the details of her turbulent existence. She had endured poverty, rape, brutal assassinations, dysfunctional marriages, and the never-ending blight of racism. But she had also reveled in motherhood, deep and uplifting friendships, and the exhilaration of unexpected experiences—including landing a singing role with *Porgy and Bess* on tour in Paris, Venice, Belgrade, Egypt, and Rome.

I Know Why the Caged Bird Sings, Angelou's debut autobiography, was published to critical acclaim in 1969, and catapulted her to literary stardom. The first nonfiction best seller by a Black woman, *Caged Bird* has never gone out of print and has become a touchstone for understanding the Black experience—most recently during the Black Lives Matter marches of 2020. "I know that not since the days of my childhood, when people in books were more real than the people one saw every day" Baldwin wrote in appreciation, "have I found myself so moved."

Angelou's genius was her voice, which she shared in disparate ways—to sing, to speak, to write. During episodes of trauma, she went silent; when it was time to be heard, she summoned it. Angelou used her voice to belt out calypso tunes and Gershwin ballads; to lecture and teach and recite powerful verses of poetry in her ambrosial cadence; and, most abidingly, to put into words the wrenching agonies and fierce joys that saturated her life and the lives of the greater Black community.

Angelou, the midlifer, never had the luxury of finding and following a passion; her journey as a writer was an unintentional odyssey borne of anguish. The first sentences she wrote introduced her to the potency of words; those that followed defied expectations. In the second half of her life, beginning in her 40s and ending with her death in her 80s, Angelou completed three books of essays, numerous volumes of poetry, and seven memoirs. Among other awards, she won three Grammys, the Langston Hughes Medal, and a Pulitzer nomination.

Angelou's voice, with its mix of intense candor and melodic pulse, is a legacy that still has much to teach us. "What I have tried to do in all of my work," she once said, "is to tell the truth and to tell it eloquently."

Marguerite Annie Johnson was born in St. Louis, Missouri, on April 4, 1928, to parents who didn't get along. Calvin Coolidge was president. Al Jolson's "Sonny Boy" was soaring on the charts. *The House at Pooh Corner* had just landed in bookstores, and the first packages of sliced bread were making their debut in supermarkets across the country.

Vivian Baxter, a nurse who later got a real estate license, met Angelou's father, Bailey Johnson, a Navy dietitian, in St. Louis in 1924. She was "the butter-colored lady with the blowback hair," as Angelou later described her; he was a soldier with a booming voice who'd served in World War I. The two were physically magnetic, but unhappily matched. "They were matches and gasoline," Angelou wrote. "They even argued about how they were to break up."

Vivian and Bailey had one thing in common, though: Neither wanted the duty of raising their children, Marguerite and her older brother, Bailey Johnson, Jr. When Bailey Jr. and Maya (her brother invented the nickname because he couldn't pronounce her name) were three and five years old, their parents shipped them off on a train with wrist tags marked with their names and destination: Stamps, Arkansas, the home of their paternal grandmother, Annie Henderson, and her disabled adult son, Willie.

Like every other town in the Jim Crow South in the 1930s, Stamps was racially polarized, with wealthy whites and poor Blacks separated by railroad tracks. As a small child, Maya dreaded the white folks with their translucent skin, small feet, and dominion over people who looked like her. The "smell of old fears, and hates, and guilt," she later wrote, mingled with the penetrating aromas of cattle dung and cured pork. "People in Stamps used to say that the whites in our town were so prejudiced that a Negro couldn't buy vanilla ice cream. Except on July Fourth. Other days he had to be satisfied with chocolate."

Angelou remembered Grandmother Henderson (she married three times, but kept her second husband's last name) as a six-foot-tall "cinnamon-colored woman with a deep, soft voice." A churchgoing disciplinarian, Momma provided warmth, direction, and much needed stability, but she was tough and didn't hesitate to wake her grandkids with a switch if they deigned to go to bed with dirty feet. When she wasn't in school, Maya spent much of her time weighing scoops of flour, sugar, and corn in Momma's country store and sitting under a neighbor's chinaberry tree.

Abandonment is difficult to reconcile at any point in life, but especially as a child. When she was little, Maya assumed her parents were dead, a presumption that spared her the pain of accepting desertion. But this illusion was shattered when one day her father, with his wide shoulders and loud laugh, showed up unannounced in Stamps. "My brother said, 'Hot dog and damn. It's him. It's our daddy,'" Angelou wrote in her memoir. "And my seven-year-old world humpty-dumptied, never to be put back together again."

Bailey Sr. had come to collect his children and take them to St. Louis to live with their maternal grandparents and subsequently their mother, who had moved back to the town where she grew up. Momma packed a shoebox of fried chicken for her grandbabies and saw them off to a place with riches they'd never known—flushing toilets, liver-wurst and salami sandwiches, and peanuts mixed with jelly beans.

In St. Louis, Maya got to know her mother's four brothers, three of whom had formed a trio she admired for their meanness. "They beat up whites and Blacks with the same abandon, and liked each other so much that they never needed to learn the art of making outside friends," she reflected. For a while, all was relatively stable. Maya got her own room and new clothes, watched her mother dance at a local pub, and spent Saturdays reading books in the public library.

But St. Louis soon scorched Maya Angelou's childhood. Her mother's live-in boyfriend, who Angelou identifies as Mr. Freeman in her autobiography, was a railroad yard foreman who brought in a dependable income and lit up when he saw her mother. The first time he sexually assaulted Maya—in the bed he shared with her mother— he threatened to kill her brother if she told anyone. Several months later, he raped her. The pain was so severe it felt like her senses had been ripped apart. "The act of rape on an eight-year-old body is a matter of the needle giving because the camel can't," she wrote.

How do we comprehend the anguish a child suffers in the after-math of this kind of trauma? It's unfathomable. Sexual abuse can lead to any number of mental health effects, from anxiety to suicide. After her mother discovered what happened, she took Maya to the hospital, where her brother begged her to tell him who did it. Maya admitted her perpetrator was Freeman, and he was tried in court. Dressed in a too-short, too-warm navy blue coat with brass buttons, Maya testified against him. Freeman was convicted and received a year's sentence but was somehow released before serving time. Three days later, he was found behind the town slaughterhouse, kicked to death.

Word was that Maya's uncles may have been involved in Freeman's demise. But Angelou conjured up another theory: Her voice—the voice that had revealed what happened—had killed him. So she shut down, refusing to speak to anyone other than Bailey, her big brother and protector. "I had sold myself to the devil and there could be no escape," she wrote in her memoir. "Just my breath, carrying my words out, might poison people and they'd curl up and die like the black fat slugs that only pretended."

Maya's muteness was accepted in the immediate aftermath of the rape. But when the doctor proclaimed her physically healed, her family's patience evaporated into the sticky St. Louis air. Maya received thrashings for having an attitude and soon after, she and Bailey were sent back to Stamps—relinquished once again. "I have never known if Momma sent for us," Angelou wrote, "or if the St. Louis family just got fed up with my grim presence."

At first, the old neighborhood seemed strange and uncomfortable. Maya couldn't remember names; colors looked faded. Eventually, as the months passed, she began feeling better. Her grandmother accepted her silence, and Maya made her first friend, a girl with braids named Louise. The two spun around and stared at the sky near the foot of a sycamore tree.

A woman named Bertha Flowers helped her recover her will to speak. One day, Mrs. Flowers invited Maya to her home and read aloud to her from *A Tale of Two Cities*. Over sweet tea cookies and lemonade, she coaxed Maya to memorize poetry, which she described as music for the human voice. It had to be read aloud, she explained. It had to emerge from a person's lips.

Poetry took Maya by the hand and led her out of silence. She learned Shakespearean sonnets, reciting "To be, or not to be" in front of a mirror. She inhaled poems by Paul Laurence Dunbar, Countee Cullen, Henry Wadsworth Longfellow, and Edgar Allan Poe, whom she affectionately nicknamed "Eap." She read Robert Benchley, Ann Petry, Thomas Wolfe, Richard Wright, and Hemingway.

Soon, her own words started to tumble. The voice she had suppressed began to find itself again.

"When I decided to speak," she said, "I had a lot to say."

⌒

Childhood shapes who we become. Shirley Temple, born in the same month and year as Maya Angelou, spent her youth loved, nurtured, admired, and celebrated. While Shirley was perfecting her tap dances with Bill Robinson, Maya was trampled by sexual abuse. As Shirley and Bojangles became the first interracial duo to dance together on-screen, Maya battled the horrors of racism. The stars aligned early for Shirley; Maya's youth was eclipsed by adversity.

When her grandchildren reached adolescence, Momma decided they needed to get out of the South. Bailey had started getting testy about the deferential behavior expected of Blacks by Stamps' downtown whites; far worse, he'd stumbled across the horrific aftermath of a lynching. There was constant fear of the Ku Klux Klan; when they came anywhere near town, the kids rushed to hide Momma's son, their uncle Willie, in a food bin, covering him with potatoes and onions so he wouldn't be discovered and falsely accused of a crime.

The only option was to send Maya and Bailey back to their mother, who by now had relocated from St. Louis to San Francisco. Yet again, they had to readjust. The disparity between southwest Arkansas and the hills of Northern California was staggering, as was the contrast between their grandmother and her unconventional mother, who wore lipstick, smoked Lucky Strikes, listened to jazz, and danced the Snake Hips. Time moved slowly in Momma's South; Vivian Baxter, by contrast, "spoke fast and she wiggled all the time," Angelou recalled.

Fully grown, Maya was six feet tall and towered over her mother, even as a teenager. She was self-conscious about her appearance, bemoaning her flat chest, thin legs, and deep voice. "That woman who looked like a movie star deserved a better-looking daughter than me," Angelou wrote.

But Vivian Baxter stepped up as a mother. She told her daughter she was beautiful and tutored her in courage. When 16-year-old Maya announced she wanted to get a job collecting tokens on the San Francisco streetcars—she especially liked the uniforms with their change belts, caps, and form-fitting jackets—her mother persuaded her to bring a book and sit in the office for as long as it took for the manager to give in and hire his first Black female employee.

It was motherhood that cemented the relationship between Maya and Vivian. After a perfunctory liaison with a neighborhood boy she barely knew, Maya got pregnant at the age of 16. The father wanted out and stopped talking to Maya during her fourth month; her mother, in contrast, welcomed the baby. "She never, ever made me feel that I had done the wrong thing," Angelou told Oprah Winfrey in a television interview years later. "She said, 'We're not going to ruin three lives. We're going to have a beautiful baby.'"

From this moment, motherhood would shape the trajectory of Maya's life. She needed to nourish, educate, and mentor the little boy she named Clyde. Lofty dreams were meant for others, enabled by the color of their skin. "The white kids were going to have a chance to become Galileos and Madame Curies and Edisons and Gauguins, and our boys (the girls weren't even in on it) would try to be Jesse Owenses and Joe Louises," she recalled. "We were maids and farmers, handymen and washerwomen, and anything higher that we aspired to was farcical and presumptuous."

Through her late teens and early 20s, Maya took work where she could get it. Some jobs were mundane (a waitress at a chicken shack); some tried her soul (she dabbled in sex work and briefly oversaw a brothel). But other positions lifted her spirits. While working as a clerk in a record store, she steeped herself in rhythms. Filled with energy and optimism, she also met and married her first husband, a handsome Greek sailor named Tosh Angelos.

Music and movement continued to appeal to Maya. She listened to Dizzy Gillespie, Sarah Vaughan, John Lee Hooker, and Charlie Parker, and swayed to music, baby in her arms; dance helped to liberate her from workaday jobs. A natural entertainer with sultry moves and a luminous smile, Angelou soon sought out the stage. In 1953, the year she turned 25, she got a gig at a San Francisco strip joint called the Garden of Allah. Her costume was sparse—a feather and a few sequins—and she got noticed. As word spread, she jumped to the more upscale Purple Onion, where she became a calypso singer, using her voice to charm her audience. To sound more distinctive as a performer, she tweaked her now ex-husband's last name (the marriage was dissolved after just a few years) from Angelos to Angelou, becoming Maya Angelou for the rest of her life.

Her big break came a year later, when a traveling tour of *Porgy and Bess* landed in San Francisco. Angelou was astonished by George Gershwin's soulful opera, set in the tenements of Charleston, South Carolina, and performed by an all-Black cast. "I had laughed and cried, exulted and mourned," she recalled. "I remained in my seat after the curtain fell and allowed people to climb over my knees to reach the aisle. I was stunned." Later, the operatic cast stopped by the Purple Onion, saw Angelou perform, and urged her to try out for their European tour. She landed the role of Ruby.

Porgy and Bess had ignited raves during its first tour in 1952; in Berlin alone, the cast earned 21 curtain calls. Angelou joined the production in September 1954 in Montreal, leaving her son, just after his ninth birthday, in the care of her mother and an aunt, Lottie, in California. Over the next eight months, she traveled to dozens of cities in Europe and the Middle East, including Marseille, Belgrade,

Casablanca, and Tel Aviv. Gondoliers serenaded her in Venice. She rode camels in Cairo. And in Paris, she met James Baldwin, the man who would later pull her from hopelessness and kick-start her life as a writer.

But as the months stretched out, Angelou became increasingly anguished about her separation from Clyde, whom she had never left for such a long period of time. One day, after seeing children playing out of a window from her seat on the Orient Express, "the longing for my own son threatened to engulf me," she recalled.

In the spring of 1955, Angelou received a letter from her mother in California telling her she was needed back home to take care of her son—and that he desperately missed her. Angelou told her producer she was going home. After sailing from Naples to New York, she made her way by train to San Francisco. "I had left my son to go gallivanting in strange countries and had enjoyed every minute," Angelou wrote, "except the times when I thought about him."

Her reunion with Clyde was heartrending, a raw emotional touch-stone in her life journey. He wrapped his arms around his mother's neck and worried that she would leave again; she despaired over neglecting him. Over the next few days, her angst escalated until she feared she was going mad.

This moment—this exceedingly painful and scary moment—led Angelou to first discover the impact of writing down words. After a failed emergency session with a psychiatrist, she knocked on the door of her old singing coach, Frederick Wilkerson. Wilkie (as she called him) gave her a Scotch, listened to her bawl over feeling guilty about leaving her son, and then handed her a yellow legal pad and a pencil. Jot down everything you have to be grateful for, he told her. "Write, dammit! I mean write," he said.

Angelou picked up the pencil and scribbled out her bare-bones list, documenting what she had to be thankful for: She could hear and talk and sing, she wrote. She knew how to dance, she could cook, and she had a family.

The anguish dissipated almost immediately. "I followed Wilkie's orders," she later reflected, "and when I reached the last line on the first page of the yellow pad, the agent of madness was routed."

⌒

Creative people in the arts—poets, writers, painters, sculptors—often emerge from troubled pasts. Psychologist Dean Keith Simonton has long studied the variables of genius and documented differences in family backgrounds. Scientific geniuses, Simonton reports, tend to come from more stable and homogenous backgrounds compared with artistic geniuses, whose parents tend to be more ethnically and religiously diverse and who are more likely to have suffered traumatic experiences in childhood or adolescence.

Numerous studies have found a connection between eminent achievers and the loss of one or both parents early in life. In one benchmark analysis launched in the 1970s, psychologist J. Marvin Eisenstadt tracked down the family backgrounds of 573 individuals who had achieved high rank in their professional occupations—starting with Homer and other ancient Greeks and Romans and ending with esteemed figures of the 20th century, including John F. Kennedy (almost all were men). Eisenstadt found that a third of the group had endured the death of a parent by the age of 15 and almost half had lost a parent by the age of 20.

Grief can trigger emotional crises, Eisenstadt noted, but in some instances it may ultimately help mold determination. "The problem of mastering a changed and changeable environment," he wrote, "can be translated into strivings for achievement, accomplishment, and power."

Physical anguish can also drive a person to discover creative outlets for their pain. Frida Kahlo, the daughter of a German immigrant father and Mexican mother, was disabled by polio at the age of six. At 18, a horrific bus accident fractured her spine and collarbone and crushed her pelvis and right foot. It also tested Kahlo's will, and remarkably, launched her artistic career. While immobilized in a full body cast, Kahlo began painting from her recovery bed. Today, she is revered as one of the most piercing and original Mexican painters.

Consider the remarkable outcomes of Jews who survived the Holocaust and grew up to receive Nobel Prizes for their professional achievements. Roald Hoffmann, who was awarded the prize for chemistry in 1981, was hidden in the attic of a schoolhouse in rural Poland during the war. Later, he managed to flee with his mother, ultimately settling in the United States; his father and other family members were murdered by the Nazis. Elie Wiesel, the acclaimed novelist and human rights activist, was liberated from Buchenwald death camp at the age of 16 with the number A-7713 tattooed on his arm. In 1986, when awarding him the Nobel Peace Prize, the Nobel Committee described him as a "messenger to mankind," stating, "His belief that the forces fighting evil in the world can be victorious is a hard-won belief."

Trauma is in no way a prerequisite for genius. But every obstacle, every challenge becomes a strand of experience woven into the whole. Who we are informs how we express ourselves. "A writer—and, I believe, generally all persons—must think that what happens to him or her is a *resource*. All things have been given to us for a purpose, and an artist must feel this more intensely," wrote Jorge Luis Borges, the Argentinian poet, essayist, and short story writer who went blind in his mid-50s. "All that happens to us—including our humiliations, our misfortunes, our embarrassments—all is given to us as raw material, as clay, so that we may shape our art."

Spark

⌒⌒

In her late 20s, Angelou began hopscotching destinations while edging toward her craft. After her meeting with Wilkie, she and her son moved to Los Angeles, where she started to write—first short sketches, then song lyrics and even short stories. There, Angelou met the novelist John Killens, a founder of the Harlem Writers Guild, who was working on a screenplay for his novel *Youngblood*. Killens read Angelou's early work, told her she had talent, and persuaded her to move to New York City to join his organization of African-American writers. In 1959, the year she turned 31, she relocated there with her son (who by then had changed his name from Clyde—"it sounded mushy," he told his mother—to Guy) and began sharing her work openly. The first few critiques stung, but she bonded with her fellow scribes, who gave the much needed perspective that writing required dogged and unremitting dedication.

Angelou's evolution in Harlem coincided with the civil rights movement. Writing was now a passion—she was thrilled to have a short story published in a Cuban magazine—but not yet a full-time pursuit. Galvanized by the quest for justice and equality for Black Americans, Angelou landed a job as coordinator for the Southern Christian Leadership Conference under the direction of Martin Luther King, Jr. The first time they met, she was surprised to discover that King was shorter and less intimidating than she expected—a powerful presence in public, but easygoing and down-to-earth in person. Looking at him, she wrote, "was like seeing a lion sitting down at my dining-room table eating a plate of mustard greens."

But it was another activist, a South African freedom fighter living in Cairo, who uprooted Angelou once again. She met Vusumzi Make at a dinner in New York and was electrified by his passion and charisma. The two married soon after, and, along with Guy, settled in

Cairo in the early 1960s, where Angelou took up a new form of writing: journalism. She wrote and edited articles for an English-language magazine, the *Arab Observer,* and scripted commentary for Radio Egypt. Words took hold of her in a new way, and she used her voice to shape opinion. When Angelou's marriage to Make ended only a year later—he turned out to be controlling and unfaithful—she and Guy decided to join a flourishing community of Black American expats in Ghana.

West Africa enthralled Maya Angelou. Ghana was prospering under a respected leader, Kwame Nkrumah, who welcomed American Blacks to his country. The expatriate community provided community and intellectual stimulation; for the first time, Angelou felt fully accepted as a person with dark skin. The Ghanaians embraced her— she looked like them!—and warmed her soul. "Their skins were the colors of my childhood cravings: peanut butter, licorice, chocolate, and caramel."

Angelou got an administrative job at the University of Ghana, where Guy, now a high school graduate, was enrolled. She also continued her work as a journalist, writing for the *Ghanaian Times* and Radio Ghana. She heard phones answered at the newspaper office in the country's rich dialects, including Twi, Ga, and Pidgin; she reveled in the musical names of city streets (Kumasi, Koforidua, Mpraeso); and she studied the Fanti language, introducing her to African sounds and tones that etched themselves into her memory and settled into her prose. "The music of the Fanti language was becoming singable to me," she wrote, "and its vocabulary was moving orderly into my brain."

That power of voice prompted Angelou to leave Africa after three years and return to the United States. After hearing the charismatic Black nationalist Malcolm X captivate an audience with his words in Ghana—he "thundered, whispered, then roared," she wrote—Ange-lou accepted his invitation to become the coordinator for his newly formed Organization of Afro-American Unity. (Guy, then 19, stayed behind.)

In February 1965, Angelou flew to San Francisco, where she intended to spend a few weeks with her mother before relocating to New York. But only days after she arrived, Angelou got word that the 39-year-old civil rights activist had been gunned down as he began to deliver a speech at the Audubon Ballroom in Harlem. The horror left Angelou depressed, angry, and unmoored.

Now at the cusp of her 37th birthday, Angelou had no clear sense of what to do next. Her brother Bailey invited her to Hawaii, where he was living, and helped her step back onto the stage as a means of moving back into joy. At the Encore nightclub in Honolulu, Angelou reclaimed her old lineup of calypso music and sang songs by Duke Ellington and the Gershwins. But the melodies that had catapulted her to *Porgy and Bess* in her 20s did not take hold in her 30s. Life in between had been filled with political and racial awakenings; more important, she realized that she could not summon the passion needed to flourish as a singer.

Determined that it was time "to reenter real life," as she described it, Angelou moved back to L.A. She hoped to complete a stage play and work on her poetry. "I had notebooks full of poems," she wrote, "maybe I'd try to finish them, polish them up, make them presentable and introduce them to a publisher and then pray a lot."

Then the Watts riots erupted. The rebellion, the desperation, the clash of poor Blacks and policemen, the contrast between Blacks in America and Blacks in Africa—all of it seeped deep inside. One night, after venturing into the thick of the downtown violence, she started writing poetry on a yellow pad at her kitchen table. In bursts of vivid language, she described the fiery blazes, the looting, and the cavernous racial tension. She turned to the rhythm of words to tame the flames of destruction.

Verse by verse, Angelou began honing her craft. In 1967, the year she turned 39, she decided it was finally time to move back to New York, the place that first nourished her literary ambition. She wrote plays, a two-act musical, and poetry. And she reunited with fellow wordsmiths at the Harlem Writers Guild, including her old friend Jimmy Baldwin, who brought her to the Feiffers' dinner that transformational evening in 1968.

After accepting editor Robert Loomis's challenge, Angelou began telling her story, unlocking her memories of childhood in language that blazes with honesty and dances with metaphors. Every word holds power; every sentence evokes raw emotion. "If growing up is painful for the Southern Black girl," she writes in her dramatic opening to *I Know Why the Caged Bird Sings*, "being aware of her displacement is the rust on the razor that threatens the throat."

Maya Angelou, at the age of 40, had found her calling.

How do we process trauma? Writing about difficult experiences, studies have shown, can act as an emotional catharsis. Journaling, for example, reduces stress, lowers blood pressure, boosts the immune system, and relieves anxiety and depression. The key is to express deep and troubled feelings, not just random musings—and the payoff can be dramatic. Researchers in New Zealand recently found that people who wrote down their emotions about upsetting or traumatic events in their lives before undergoing a skin biopsy healed faster than patients who wrote about unemotional activities—perhaps because letting out deep angst allows for better sleep and may even reduce stress hormones that can interfere with recovery.

For some writers, autobiography is a gateway to recovery, a place where truths are unsealed and revealed. Among American Black

Spark

writers, the genre dates back to powerful accounts written by slaves. In his 1845 narrative, Frederick Douglass documented the heinous behaviors of the slave masters he encountered and the horror he felt after witnessing a whipping for the first time as a child. "It was the blood-stained gate, the entrance to the hell of slavery, through which I was about to pass," he wrote. "It was a most terrible spectacle."

Two decades later, a slave named Elizabeth Keckley chronicled her traumas—her separation from her father, whom another overseer owned, the vicious beatings, and sexual assaults. Keckley's narrative also included memories of the four years she lived in the White House as Mary Todd Lincoln's dressmaker and close confidante, where she witnessed the aftermath of Lincoln's assassination: "I shall never forget the scene—the wails of a broken heart, the unearthly shrieks, the terrible convulsions, the wild, tempestuous outbursts of grief from the soul."

These works of Black writers—slave narratives as well as poetry, spirituals, and essays that followed—appealed to Angelou. By its very nature, autobiography necessitates vulnerability, and Angelou cherished the challenge. "It's allowing oneself to be hypnotized. That's frightening, because then we have no defenses, nothing," she said. "We've slipped down the well and every side is slippery. And how on earth are you going to come out? That's scary. But I've chosen it, and I've chosen this mode as my mode."

Angelou grounded her writing in candor and painted her stories in sweeps of rich color. She was at times brutally frank: Reverend Howard Thomas, from her Arkansas childhood, is "ugly, fat, and he laughed like a hog with the colic." Two lesbian women, who try to seduce her when she is in her late teens, are "thick-headed, lecherous old hags." But there is so much majesty in her imagery. A man's smile "showed teeth so anxious they clambered over each other." A fellow performer in *Porgy and Bess* has a voice so flawless it is "like hot silver being poured from a high place."

When it was published in 1969, *I Know Why the Caged Bird Sings* brought Angelou recognition and acclaim. The writing emerges "right from the center of the blackness," wrote Ward Just in the *Washington Post*. It was her story, Angelou noted, but she intended to speak to the human condition. "I'm trying to say in that book you can win," she said in a 1970 radio interview with Studs Terkel. "I'm saying it to young black children. I'm saying it to old black men and women. To middle-aged Chinese. Teenage whites. I'm saying it about the human condition, really."

Over the next 44 years, Angelou completed six more memoirs. The final installment, *Mom & Me & Mom,* was published in 2013, the year she turned 85.

There was method to her craft. Angelou rented a hotel room near her home, where she worked from about 6:30 in the morning until 1:00 or 2:00 in the afternoon. With no distractions—she insisted that artwork on the walls be removed—she wrote longhand using a ballpoint pen on the same kind of yellow legal pad Wilkie had provided at that milestone moment earlier in her life.

Her accoutrements included a deck of playing cards (solitaire helped entertain her "little mind" so her "big mind" could get to work); a bottle of sherry (a glass around 11 a.m. helped the muse); and a trio of literary aids: Random House's dictionary, *Roget's Thesaurus,* and the King James Bible. A desk was unnecessary; she lay across the bed, digging her elbow into the mattress so deeply it was covered in callouses.

Drafts of Angelou's writing show a dignified looping cursive, with words scratched out and added. Her sentences flowed because she labored tirelessly, not because the inspiration came easily. In fact, she bristled when people called her a natural writer. "Those are the ones I want to grab by the throat and wrestle to the floor," she once said, "because it takes me forever to get it to sing. I *work* at the language."

Spark

At times, she got lost in a pattern of self-doubt and felt like a fraud. "I get caught up in my insecurity despite the prior accolades," she wrote. "I think, uh, uh, now they will know I am a charlatan that I really cannot write and write really well." While writing her fourth memoir, *The Heart of a Woman,* she recalled moments when her confidence evaporated like smoke wafting through a room. "It would just dissipate and I would suddenly be edgewalking again."

Still, Angelou pushed through and learned to bury the notion that every piece had to attain some kind of mythical perfection, a masterpiece that dazzled like the Sistine Chapel. That burden was especially heavy for Black writers, she said, because they had more to prove. She confessed in an interview with the *Paris Review* that she wrote until she had done the best she could do. "I know that one of the great arts that the writer develops is the art of saying, 'No. No, I'm finished. Bye.' And leaving it alone," she said. "I will not write it into the ground. I will not write the life out of it."

Throughout her life, Maya Angelou was embracing of people and things. She had a hankering for Hershey's Kisses and Hebrew National hot dogs, loved lemon chicken and lemon pies, and savored every bite of the canned pineapple her grandmother gave her for Christmas. She delighted in gathering friends and family around her table for a home-cooked meal and held strongly to her faith. "I am a child of God," she told Oprah.

In 1982, when she was 54 years old, Angelou was named a professor of American Studies at Wake Forest University in Winston-Salem, North Carolina, where she taught for the last three decades of her life. She was an alluring educator, captivating her students with her stories, and cooking for them at her house. She was electric, one of her students later recalled in a dedication in *Wake Forest Magazine.* "I felt the air change when Maya Angelou entered the room."

During the second half of her life, Angelou collected experiences as she always had, like stones along a riverbank. In 1973, when she was in her mid-40s, she appeared in a two-person Broadway drama about Mary Todd Lincoln—playing the role of Elizabeth Keckley—and got married for the third time, this time to the younger Paul du Feu, the former husband of author Germaine Greer. The two met at a Soho literary event. "This tall, handsome Englishman came up and asked if I were alone," she said in an interview in 1975. "I said, 'Yes, why?' and he told me I was the most beautiful woman in the world, and could he take me to dinner—right now?" The marriage was gratifying for a while—they renewed their vows twice—but ended in divorce after eight years.

Once, when somebody asked her about aging, Angelou replied, "Do it. You don't have much of a choice." It was how she led the latter decades of her life. She played the role of Kunta Kinte's grandmother, Nyo Boto, in Alex Haley's 1977 television series, *Roots,* for which she received her Emmy nomination. She wrote verses for Hallmark cards and hosted a radio show. At age 70, she directed her first feature-length film, *Down in the Delta.*

On January 20, 1993, Maya Angelou, wearing gold hoop earrings and a dark blue coat with brass buttons, stood at a podium on the West Front of the U.S. Capitol: the second poet and the first African American and woman to read a poem at a presidential inauguration. Just after President William Jefferson Clinton's swearing in, she stepped to the microphone, greeted the crowds in the winter sunshine, and began reciting, "On the Pulse of Morning," the poem she had penned for the occasion. In the six minutes it took Angelou to read her words, she described America's beauty and the arc of its troubled history. She honored her country's rich diversity and gathered its citizens—from Sioux to Sheikh—together in verse. She chose hope and courage over pain and fear.

On May 28, 2014, Maya Angelou died at the age of 86 at her home in Winston-Salem. Toward the end of her life, she had received the Presidential Medal of Freedom from President Obama, and she was posthumously awarded a Congressional Gold Medal for her contributions to American culture and the civil rights movement. Atop her many remarkable achievements, she was a mother, a grandmother, and a great-grandmother. At her funeral, her son Guy said, "There is no mourning here. There *is* no mourning. We have added to the population of angels. And she has left each one of us with something in our heart."

And so she did. Angelou's writing influenced countless writers, including Toni Morrison and Alice Walker. Tennis champion Serena Williams recited Angelou's poem, "Still I Rise," at an acceptance speech for Sportsperson of the Year; Meghan Markle quoted Angelou in a speech she gave in South Africa. Angelou guided Michelle Obama with her confidence, calm, and grace, and became a mentor and cherished friend to Oprah Winfrey, who called her "the greatest woman I have ever known."

Some people progress through life, one foot in front of the other. Maya Angelou danced, fell, leaped, tripped, and soared—grasping every opportunity and every failure as vital parts of her voyage. Her voice was her bequest to humanity. "It is not like she woke up one day and decided to be Maya Angelou. It is the conglomeration of all of these different experiences and aggregating them together," her grandson Colin Ashanti Murphy-Johnson said after she died. "And the aggregate is Maya Angelou."

CHAPTER 9

Alexander Fleming

The Microbiologist (1881–1955)

O ne autumn morning, I make my way down Praed Street in
the heart of London. The neighborhood, home to Padding-
ton Railway Station, is lined with cafés and shops—a
pawnbroker, a Western Union currency exchange, a hairdresser—and
bustling with tourists and city dwellers toing and froing.

My destination is St. Mary's Hospital, which although far from
posh-looking, has deep ties to British royalty. Queen Elizabeth the
Queen Mother served as president of the hospital from 1930 until
her death in 2002, and a slew of royal births have taken place here,
including Prince William in 1982, Prince Harry in 1984, and, most
recently, Prince William and Kate Middleton's children: Prince
George, Princess Charlotte, and Prince Louis.

My visit has nothing to do with the monarchy, however. I have
come to see the birthplace of one of the greatest triumphs in modern
medicine. Here, one late summer day in 1928, an unassuming scientist

named Alexander Fleming discovered penicillin—a breakthrough that would propel Fleming to fame, save the lives of millions of people, and change the course of medical history.

Unlike Albert Einstein, Paul Dirac, or Isaac Newton, whose revolutionary theories about time, space, and gravity surged to the fore when they were in their 20s, Fleming did everything later in life. He started medical school two years after his peers, married in his mid-30s, became a father in his early 40s, and at the age of 47 had his monumental lightbulb moment: The mysterious mold growing in his petri dish had the potential to kill off toxic germs.

Over the two and half decades that followed, Fleming morphed from a dogged lab scientist into an acclaimed public figure knighted by a king. In 1945, at the age of 64, he won the Nobel Prize in Physiology or Medicine. In his acceptance speech at the Nobel Banquet in Stockholm, Fleming stressed that he had been blessed with good fortune—not just because penicillin dropped into his lab dish by chance, but also because his position at St. Mary's allowed him to put his other research aside and pursue it—to "follow the track which fate had indicated for me."

Throughout his life, Fleming was a modest man who viewed his breakthrough as kismet more than strategy. "It may be that while we think we are masters of the situation," he said in his speech, "we are merely pawns being moved about on the board of life by some superior power."

But there was far more to the story. Although luck clearly played a role, Fleming's years of preparation allowed him to deliver the goods. A younger scientist would not have had the decades of experience and deep knowledge of science that bolstered Fleming's skills—his acute power of observation, his broad manner of thinking, his unwavering patience, and his willingness to follow a lead without anticipating where it might take him.

Fleming was the ideal scientific pawn precisely because of where he was on his life trajectory: a midlifer at the cusp of late blooming. In medicine, especially, discoveries happen when scientists have the know-how to make them. In the famous words of Fleming's predecessor, the French microbiologist and chemist Louis Pasteur: "Chance favors the prepared mind."

⌒

Alexander Fleming was born on August 6, 1881, on Lochfield Farm in Ayrshire, Scotland—a quiet place in a world beset by turmoil. Within weeks of Fleming's birth, U.S. president James Garfield would succumb to an assassin's bullet, the Warsaw pogrom would assail Russian-controlled Poland, and ferocious Typhoon Haiphong would kill thousands of people in Southeast Asia.

Alongside these catastrophes came the perilous onslaught of disease. Outbreaks of smallpox, typhus, yellow fever, and cholera had killed millions in the 19th century; the deadly 1918 influenza, transmitted by coughs and sneezes, loomed on humanity's horizon and would influence Fleming's career. Before it finally dissipated in 1920, the flu had killed more than 50 million people worldwide—many of them young adults under the age of 40.

Lochfield's 800 acres provided a bucolic retreat from the outside world. Alexander's father, Hugh, was a sheep farmer and widower with four young children when he married Fleming's mother, Grace Morton, in 1876. Over the next six years, the couple had four children together: a daughter, also named Grace, followed by three boys, John, Alexander, and Robert.

The farm bubbled with adventure—especially during the warm summer months, with Fleming children rambling through the fields, rolling down the hills, splashing in the streams, and collecting eggs, which they sold to a local grocer. Alec, as Alexander was known in his family, was especially close to his younger brother, Robert, and the two would remain devoted siblings throughout their lives, hunting together for rabbits with their bare hands, fishing for trout, and becoming adept bird-watchers who could identify hawks, plovers, swallows, and gulls.

Like Charles Darwin, who was smitten by the outdoors as a child, Fleming was a born naturalist. His early pastimes on the farm fostered his inquisitiveness about the world around him—a skill that he credited to his childhood later in life. "My powers of observation were sharpened by my search for peewits' eggs in the fields and moors," he recalled, "my patience increased by the guddling for the trout in the Glen Water."

The Fleming family had its share of upheaval. Alec was just seven years old when his father died after being incapacitated by a stroke a year prior. Grace Fleming kept the children together after her husband's death, playing games with them and keeping them safe. On cold winter mornings, she gave her children two warm potatoes, which served a double purpose: warming their hands during the long walk to school and providing a meal. Fleming later praised her for a job well done. "I have to render great thanks to my mother—one of the best women who ever lived—for my upbringing," he said.

An avid reader and lucid thinker, Fleming routinely outperformed his fellow students at school. "He could extract the essential facts from a book or a lesson, tuck them away in his memory and recall them at will, even years later," noted his biographer Gwyn Macfarlane. This might have prepared Alec for any number of professions, but a simple change in geography helped steer his early career. When he was a young teenager, Alec moved to London to join three of his four older half siblings who had relocated to the city—including Tom, who encouraged his brother to pursue business, a career he believed would provide Alec with financial security.

Alexander Fleming

Sometimes, knowing what we don't enjoy leads us closer to finding what we do. Accepting Tom's advice, Alec quit the Regent Street Polytechnic at the age of 16 and took a job as a junior clerk for the America Line, a shipping company that transported goods between England and the United States. Confined mostly to the office, Fleming earned 10 shillings a week for copying documents, keeping track of accounts, and preparing passenger manifests. "The only problem," writes Kevin Brown in his biography, *Penicillin Man*, "was that he hated the job."

Just as World War II propelled Julia Child's career, the 1899 Boer War provided Fleming with a purpose. The British Army needed volunteers to maintain Great Britain's dominion over the two Boer Republics in South Africa. Nineteen-year-old Fleming joined the London Regiment, mostly men of Scottish descent, finding a welcome diversion outside of his mundane clerkship.

Fleming's service provided him with skills that would later lead him to St. Mary's for his medical studies. As a member of the regiment's "H" company, he underwent part-time military training and participated in marches. His experience earned him wartime service and provided a social outlet—a place where he could rendezvous with fellow Scots and compete in sporting events, including water polo and rifle shot.

In 1901, a bout of good fortune materialized. A bachelor uncle left Fleming a generous sum of £250 (around £31,000, or $40,000, today). No longer dependent on a paycheck, Fleming could now search for a better-suited career. By this time, Tom had established a successful ophthalmology practice and steered his younger brother to medicine. Fleming quit his job, took the required coursework, did exceptionally well, and earned his pick of a dozen medical schools.

Fleming chose St. Mary's Hospital Medical School based on little more than a youthful observation: He remembered playing water polo for the military regiment against St. Mary's medical team and fondly recalled the lively camaraderie among teammates. With little knowledge about how one school compared to the other, his memory of that sports match clinched his choice, writes Brown, and "it was to be a decision that neither Fleming nor St. Mary's was to regret."

Unlike prodigies, whose early passions set the stage for advanced performance early in life, Fleming never seemed to fret about the pace of his achievements or his age. He was 20 when he entered St. Mary's with a scholarship in the fall of 1901—two years older than most of his medical school classmates—but he considered his experience beneficial. His clerkship had not provided academic training, but "I gained much general knowledge," he said, according to his biographer André Maurois, "and when I went to the medical school I had a great advantage over my fellow students who went straight from school and had never got away from their books into the school of life."

Fleming had another advantage: He was brilliant. At St. Mary's, he aced his studies and racked up prizes and class awards, including best medical student. After passing two demanding exams required for a surgical fellowship, he seemed destined to become a surgeon, which would have likely meant leaving St. Mary's for training elsewhere.

But once again, fate intervened in the guise of sports. Hearing that Fleming was a first-rate shot, a young rifleman and bacteriologist named John Freeman hatched a plan to keep Fleming at St. Mary's so that he could join their rifle team to bolster its chance of winning an upcoming competition. Freeman talked Fleming up to his boss, Sir Almroth Wright, who agreed to offer the medical school graduate a position in the inoculation department. Here, the big spark of discovery would take place.

Even when life trajectories are planned out, improbable events crop up to change the course of history. Mary Shelley might never have written *Frankenstein* if it were not for the explosion of Indonesia's Mount Tambora in 1815. The largest volcanic eruption in recorded history covered the earth in a haze that led to the "year without a summer" of 1816—prompting Mary to hole up with her future husband, Percy Bysshe Shelley (the two would marry later that year), the writer Lord Byron, and other companions at Villa Diodati on Lake Geneva. Had the sun been shining, the group would have been outside boating and hiking. But the unusually cold, wet, and gloomy weather forced them to shelter inside, where Byron challenged everyone to write a ghost story to pass the time. Bad weather begat Frankenstein.

It is mind-boggling to imagine the what-ifs in Fleming's life. If he'd had poor aim as a rifleman, he would have likely pursued his intended path to surgery. Had he become a surgeon, he would almost certainly never have discovered penicillin. Had he not discovered penicillin, modern medical history would have looked very different. You and I might not have been born if our parents or grandparents had succumbed to incurable infections before Fleming intervened.

In St. Mary's inoculation department, Fleming found himself captivated by the research and by Wright, the famed bacteriologist whose courses he had attended as a medical student. The two made an interesting pair. Fleming, at about 5 feet 7 inches, "was a very humble figure, the 'little man' as he was affectionately called," Freeman remembered. He had broad shoulders, white hair, piercing blue eyes, and a badly flattened nose, the result of a childhood collision with a schoolmate.

Wright, by contrast, stood six feet-plus, sporting eyebrows so demonstrative they could "almost speak," Freeman noted. Opinionated, stubborn, and an ardent anti-suffragist, Wright believed that women had no place in medicine—or, for that matter, in his lab. And

Spark

he allowed little room for dissent when it came to his strong conviction that vaccines would work well as therapeutic medicine, a theory for which there was little definitive evidence at the time. Critics nicknamed him "Sir Almost Wright." Photographs show him brooding.

Despite their differences, Wright and Fleming developed a loyal scientific and intellectual partnership that would last for four decades. Among other attributes, Wright had a powerful group of friends who supported his lab when he needed financial assistance. And in addition to his scientific brainpower, he charmed Fleming with his passion for philosophy and poetry. Visitors to Wright's lab included statesmen Arthur Balfour and John Burns, and dramatists George Bernard Shaw and Harley Granville-Barker. In one of Wright's favorite pastimes, he recited poetry during teatime, after which he quizzed his audience about the writer's identity. Was it Shakespeare? Wordsworth? Kipling? Fleming amused the lot of them by answering with the same name every time: "Robbie Burns," a fellow Scotsman.

Fleming entered medicine at a propitious time. In the mid-1800s, Louis Pasteur's research led to the groundbreaking revelation of germ theory, which held that certain diseases are caused by tiny microbes that assault the body. Pasteur's work, which included developing vaccines for anthrax in animals and rabies in humans, set the stage for Fleming's discovery of penicillin, a mold that could combat pernicious bacteria.

The battlefields of World War I catapulted Fleming's work. In the fall of 1914, several years into his tenure, he and fellow scientists traveled to Boulogne, France, where Wright set up a lab in a converted casino to treat war wounds. The team quickly discovered that soldiers were suffering from severe septic infections connected to their injuries, for which the doctors could do very little—an experience that motivated Fleming to look for answers. "It made me especially interested in substances that killed microbes," he later said.

Fleming's wartime service provided the scientist with another deadly assault to put under his microscope: the influenza pandemic that surged in the fall of 1918. In Boulogne, Fleming examined samples from infected soldiers. What caused the flu? he wondered. And why was the pathogen so deadly? It would be years before microbiologists would have the knowledge and tools to isolate and fully understand the virus, but in the meantime, Fleming figured out that many of the deaths were due to secondary bacterial infections. He knew what he had to do: Experiment, study, and learn.

Back home at St. Mary's, Fleming immersed himself in his research. Fellow scientists remembered seeing him bent over the lab table in deep thought, often with a cigarette dangling from his lips. Curiosity drove Fleming's quest, and the more he investigated and studied, the deeper his understanding of bacteriology grew. "It was said that he knew his bacteria," an obituary in the *British Medical Journal* would later note, "as a shepherd knew his sheep."

Creativity, a feature usually associated with writing, music, and art, is critical in science; it drives the way researchers like Fleming dream up and design their experiments. One day, in his quest to better understand bacteria, Fleming decided to investigate the biological components of his stuffed-up nose by adding a dollop of his own mucus to a lab dish and setting it aside.

A number of days later, Fleming noticed something unusual: The drippings from his nose appeared to be inhibiting the growth of a certain type of bacteria he had prepared in a lab dish. Fleming showed it to his colleague, V. D. Allison, who later recalled Fleming's understated comment about what he'd found: "This is interesting."

This discovery made Fleming wonder: Could other bodily fluids ward off germs, too? Tears, perhaps? Allison and other colleagues agreed to collaborate in his research. First, they needed to make their eyes water. They tried onions, but found that lemons worked better.

Spark

"We used to cut a small piece of lemon peel, squeeze it in front of an eye while looking in the mirror of a microscope," Allison later recalled. Even lab attendants, paid three pence each, were asked to shed a few drops for the sake of science.

Like nasal mucus, tears were found to be potent at dissolving some bacteria. Fleming discovered that the bodily fluid was ubiquitous—in hair and fingernails and in plants and animals, too. He named the active substance in both of them "lysozyme" ("lysed," because it dissolved certain microbes and "zyme" because it was determined to be an enzyme).

Fleming's research proved monumental in the field of bacteriology, confirming that the body had a natural defense system able to fend off certain kinds of infection. There was only one problem: Lysozyme wasn't strong enough to conquer the most pathogenic bacteria, including typhoid, cholera, and dysentery.

Fleming needed to find something far more powerful—a natural substance that could quash virulent and fatal diseases.

⌐⌐

Alexander Fleming was a scientist with a life outside the lab. Like James Watson and Francis Crick, who frequented their neighborhood pub, the Eagle, Fleming liked to relax with a drink. After his workday was done at St. Mary's, he often dropped in at the Chelsea Arts Club, a gathering place for artists, for a drink and a game of snooker.

Fleming had earned a spot at the Arts Club after submitting a painting, a hobby he took on outside the lab. He had an eye for beauty—a talent that paid off for Allison when Fleming persuaded him to bid three pounds for several unframed etchings by an artist named Picasso, at a Sotheby's auction in 1924. His own work included watercolors and what became known as germ art. Using the natural hues of microbes as his colors—yellow from *Staphylococcus,* red from *Bacillus prodigious,* blue from *Bacillus violaceus*—he meticulously crafted landscapes and portraits.

Fleming's extracurriculars enriched his life, as did the company of his family. The scientist had found a partner in Sarah McElroy, an Irish nurse who went by Sally and later adopted the name Sareen. An extrovert, Sareen was a charmer "who could speak and laugh for both of them," notes biographer Gwyn Macfarlane, but was also organized and practical. The two were married the day before Christmas Eve in 1915; nine years later, Sareen gave birth to their only child, a son named Robert, when Fleming was 42 years old.

On the weekends, the Flemings motored to their red-tiled-roof refuge near the river in Barton Mills, a village about 80 miles north of London. With its six bedrooms, three living rooms, and a garden, "the Dhoon," as it was known, became a frequent gathering place for family and friends. Fleming especially loved boating, fishing for perch and pike, growing strawberries and tomatoes, and shopping for antiques.

Just as Newton's rural home in Woolsthorpe provided a sanctuary for his research during the plague, the Dhoon played a major role in the scientist's discovery of penicillin. More than just a beloved getaway, it was the very place to which Fleming retreated in the late summer of 1928 for his holiday, as he always did. Only this time, he had left something important behind on his worktable.

Unlike the revelations that emerged during Newton's annus mirabilis, Fleming's discovery did not occur during his time away; it stewed in his lab and waited for him to come back and notice.

⌒

Among Fleming's many notable traits, one in particular proved to be his greatest asset: He was untidy. Allison, who worked with Fleming

Spark

in his lab for five years and became a lifelong friend, remembered the stark contrast between his workspace and his colleague's. "At the end of each day's work, I cleaned my bench, put it in order for the next day and discarded tubes and cultures plates for which I had no further use," he recalled. Fleming, meantime, teased Allison about compulsion for neatness and left his own containers out "for two or three weeks until his bench was overcrowded with 40 or 50 cultures."

Fleming's disorderliness turned out to be a boon for science. Before departing for his six-week summer respite, Fleming left several petri dishes containing colonies of staphylococci, stubborn and potent bacteria he was studying, on his table. The Alexander Fleming Laboratory Museum has re-created the scientist's lab as it looked the day he returned to work, and my trip to St. Mary's allowed me the opportunity to relive it. An image of Fleming's famous lab dish guided me to a small wooden portal inside the hospital gates. As the door shut behind me, I imagined Fleming climbing the same narrow staircase to his lab almost a century earlier, knowing nothing about what he was soon to discover.

Kevin Brown, Fleming's biographer and the museum's formidably knowledgeable curator, guides me to the scientist's 12-foot-square lab on the second floor. It is filled with lab dishes, test tubes, an incubator, a Bunsen burner, the first microscope Fleming owned, handwritten text of a broadcast he made to troops in the Pacific in 1945, and an original copy of the *Times,* dated September 3, 1928—the day Fleming returned to the lab from his holiday.

That day, as he examined the dishes on his worktable, Fleming noticed something highly unusual: A foreign substance had made its way into one of the containers and inhibited the growth of *Staphylococcus aureus* microbes. Ever understated, his first words echoed the reaction he'd had to lysozyme several years earlier: "That's funny."

Later, Fleming described precisely what he found: "A mould spore, coming from I don't know where, dropped on the plate. That didn't excite me, I had often seen such contamination before," he said. "But what I had never seen was staphylococci undergoing lysis around the contaminating colony. Obviously something extraordinary was happening."

A photograph that Fleming took of the lab dish (the original is now locked away in a safe in the British Library) shows a cluster of fluffy white mold surrounded by a distinct zone in which staphylococci have begun to break down—the process of "lysis" that Fleming described. "[A]ll round the mould, the colonies of staphylococci had been dissolved and, instead of forming opaque yellow masses, looked like drops of dew," biographer Maurois noted. The bacteria farthest away from the powerful substance, meanwhile, appeared unchanged. Looking at the image, it reminded me of a castle (the mold) with a moat (the zone of disintegrated bacteria): Wade into the water and you will be attacked by castle guards.

The idea that mold might contain antimicrobial properties that could destroy bacteria was not new. A researcher at the Pasteur Institute in Paris had treated a group of tuberculosis patients with an extract made from a mold called *Aspergillus fumigatus,* and though his results were not conclusive, its potential as a remedy excited scientists. Other researchers claimed to have tested molds against different forms of bacteria as well—but none had yet succeeded in isolating and manufacturing an ideal candidate for medical use.

With a scalpel, Fleming picked off a piece of his unknown contaminant, and placed it into a tube of animal broth, used as a nutrient to grow the mold. Within a few days, the center of the mold had turned dark green and the liquid had become bright yellow, forming what Fleming initially referred to as "mould juice." The scientist cultivated his juice in lab dishes and tested the substance successfully against other bacteria, including microbes that cause gonorrhea, scarlet fever, and strep throat. A colleague helped Fleming identify it as belonging to the

Spark

Penicillium genus, first described in a scientific report in 1809, prompting Fleming to christen his new substance penicillin.

What is so fascinating about this sequence of events is that the mold might never have grown if Fleming had been a neatnik. Not only was his lab table messy but it was also unsterile, allowing contaminants to float through the air and land wherever they pleased. S. J. Woolf, an American journalist and artist who visited Fleming at his lab years later, noted the look of the place. "Here was made one of the great discoveries of modern bacteriology, yet the entire place seemed more a part of the Victorian era," Woolf reflected. "There was little about it of the sterilized sanctums of American researchers."

This was by no means an accident. Fleming spoke openly about his disdain of showy labs with gleaming marble halls. What mattered wasn't the look of the place, he insisted, but the kind of research that was being performed inside. Fleming made this point during a visit to the United States in 1945. "Had my laboratory been as up to date and as sterile as those that I have visited here," he said, "it is possible that I would never have run across penicillin."

Just as the climate precipitated Mary Shelley's *Frankenstein,* it played a starring role in the penicillin breakthrough and the antibiotic era that followed—and in a most remarkable way, Brown explained to me. Penicillin mold needs cool weather to multiply and grow; records for the summer of 1928 in London show that it was unusually chilly for nine days in late July and early August. Germs, on the other hand, rely on warmer temperatures. As if orchestrated by some meteorological conductor, balmy weather followed in mid-August, allowing the germs to propagate.

In addition to the bizarre weather, the messy lab table, the right bacteria, and the perfect mold, Fleming had something equally important going for him: middle age. Remember Albert-László

Barabási's study, which found that big hits can come at any time? He and his team noted numerous scientific breakthroughs that transpired midcareer. "So if you missed that early spark, don't despair," he wrote.

This midlife pinnacle appears to be trending older. Researchers Benjamin Jones at Northwestern University and Bruce Weinberg at Ohio State University analyzed 525 Nobel Prize winners in physics, chemistry, and medicine between 1901 and 2008. Jones and Weinberg documented the age at which the winners produced their award-winning work and divided them into two time periods: early (before 1905) and late (after 1985).

The researchers found that in the early period the majority of great discoveries occurred before the age of 40 and some 20 percent transpired before the age of 30—a time line that became almost nonexistent by the end of the 20th century. In medicine, Fleming's field, the average age of prize-winning work increased significantly, from 37.6 years in the early period to 45 in the late.

Scientist Arthur Ashkin recently took this trend to an even more impressive level. After more than 15 years of experimentation in his lab in the 1970s and '80s, Ashkin invented "optical tweezers" out of laser beams. His breakthrough moment came in 1986 at the age of 64, when he figured out how to use these tweezers to pick up viruses and bacteria without damaging them.

In 2018, when he was 96 years old, Ashkin received the Nobel Prize in Physics, earning him the honor of being the oldest laureate in history—until 97-year-old chemist John Goodenough surpassed him a year later. "So I just about made it, huh?" Ashkin joked in an interview after he received his 5 a.m. phone call from the Nobel Committee. "Because you can't be dead and win."

What might account for these later aha moments? Jones and Weinberg suggest several possibilities. For one, there are practical concerns. More experienced scientists with solid reputations are more likely to receive funding than newcomers with no track record, allowing them to focus more intently on their research.

They may have also been granted the luxury of working independently, delving into an area of interest and following intriguing leads wherever they went. This was certainly true for Fleming, whose close relationship with Wright gave him the independence to pursue his hypothesis—a luxury that many scientists are not afforded today.

When Jones and Weinberg compared early breakthrough moments with later ones, they also discovered another explanation: Scientists who are younger when they make their discoveries tend to take leaps, proposing radical theories. Take Werner Heisenberg, the theoretical physicist who published his theory of quantum mechanics when he was just 23 years old, not long after almost failing his doctoral exams. Heisenberg didn't rely on classical physics to generate his theory; his conceptual idea evolved out of seeing the field in a whole new way.

Others, however, tend to slog. Charles Darwin spent two decades perfecting his theory of evolution because he relied on established science, as well as painstaking research, to inform his proposition that humans evolved from animals. When Darwin published his opus, *On the Origin of Species,* in November 1859, he was 50 years old and the father of seven children, the eldest of whom was already 19 years old (three others had died in infancy or childhood).

Fleming, the 47-year-old, fell into the Darwinian camp. The day he discovered the mold that would be transformed into the medicine penicillin, he was at the two-thirds mark in what would be his 73-year-long life. Without his years of work on lysozyme and other substances over the prior two decades, he might never have taken note of the contaminant in his petri dish, nor thought carefully enough about what it might mean. A scientist more dedicated to color variation in staphylococci—a focus of researchers at the time—might have missed the contaminant entirely and tossed out the dish. Fleming's hunch, on the other hand, screamed out: Pay attention!

Fleming's discovery emerged from preparation. "The discovery of the powers of penicillin did not occur as a lightning flash of genius, but the way up to it had been built by long years of painstaking observation," his colleague Freeman later wrote, noting that Fleming's specific variety of mold was found to be *Penicillium notatum*. "He never tired of showing us culture plates showing the action of one colony of microbes in enhancing or retarding the growth of some neighboring colony. Thus the entry of *Penicillium notatum* onto the stage had been systematically built up. Fleming was, so to speak, waiting for it."

Science is like baseball. The game seems simple at first glance: Hit the ball, run to first base then to second to third to home. But as any player knows, there are more strikeouts than dingers. And unpredictability—a wild pitch or a stolen base—can turn the game upside down. In a similar fashion, published reports about scientific discoveries are presented in a logical and ordered way; they have to be, or they wouldn't get published.

But what often appears a flawless process is almost always filled with unexpected mishaps—as well as lucky breaks—along the way. And Fleming knew better than anyone that serendipity played a vital role in his work; he later compared the chance meeting of mold and bacteria in his lab dish to "winning the Irish Sweepstakes."

The same may be said of many of history's most notable discoveries. The poisoning of troops with mustard gas during World War I led to the discovery that the chemical slashed white blood cell counts—

a positive outcome for leukemia and lymphoma patients—and led to the production of chemotherapy. In 1879, Johns Hopkins chemist Constantin Fahlberg discovered during dinner that a chemical he'd been handling in the lab had spilled on his hands and had a very sweet taste; he developed the compound into saccharin.

Even the most unlikely scientific inventions evolve from happenstance. After returning home from a hunting trip with his dog in 1948, Swiss engineer George de Mestral noticed that his clothing and his dog's fur were covered in burrs. Upon careful examination, he discovered that the burrs had tiny hooks, which led him to the revolutionary innovation we know as Velcro (a combination of the French words *velours* (velvet) and *crochet* (hook).

These plotlines are important reminders that scientific breakthroughs like Fleming's are very often a mix of hard work and good fortune. Knowing this humanizes genius. This impressed me especially on a visit to the Nobel Prize Museum in Stockholm. There, I learned the remarkable story of Bell Labs scientists Arno Penzias and Robert Wilson, who in 1964 set out to analyze radio waves from the Milky Way on a horn-shaped antenna in Holmdel, New Jersey.

The researchers were stymied by a static noise that wouldn't go away. Thinking the sound might be interference from New York City, they directed their antenna toward the city and then away from it to see if it made a difference. It did not. Could it be coming from a pair of pigeons that had nested inside the antenna? A shotgun took care of that possibility, but didn't solve the problem. ("It's not something I'm happy about, but that seemed like the only way out of our dilemma," Penzias said later.)

For almost a year, Penzias and Wilson wrestled with the mystery until one day they heard about evolving research on the big bang, which postulated that if the universe had indeed started with a violent explosion and expanded over time, there would be low-level radiation found throughout the universe. The scientists soon realized that the noise they were hearing was neither city din nor birds; it was precisely this background radiation left over from the formation of the universe some 13.8 billion years earlier. In 1978, Penzias and Wilson were awarded the Nobel Prize for their discovery. "It's rather amazing, looking back at it, that we were looking for one thing," Wilson reflected, but "found something very much more important than what we were looking for."

For every one of these stories there are hundreds more, many of which tend to have one thing in common: a scientist's confidence in his work and an open-mindedness about where it might go. Even though Alexander Fleming didn't know what would come of the moldy substance in his dish, he was captivated by the possibilities. Years later, his colleagues remembered the moment vividly, says Brown. But at the time, they thought, "'Oh well, this is just another quirky thing Fleming's found.'"

New ideas take time to take hold. The late Dr. Judah Folkman, a wise and gentle soul, told me when I interviewed him for *Newsweek* magazine years ago that when he first proposed his then radical theory of angiogenesis at scientific meetings in the 1970s—that cancer tumors grow by recruiting blood vessels for nourishment—researchers laughed in the corner of the room or excused themselves to go to the bathroom. Today, angiogenesis drugs are helping to ward off malignancies and extend life for patients. "You have to think ahead," Folkman said. "Science goes where you imagine it."

Fleming imagined it. And, like Folkman, he didn't tie his ego to his work. If anything, he played it down. Fleming's diffident demeanor made some people uncomfortable, Brown says—especially when he'd stare at them without saying a word. He maintained an air of mystery, even with his closest colleagues. "His contribution to any laboratory discussion was usually in the form of Scotch grunts and terse inter-jections, or he reserved his words till others had talked themselves out," his colleague Freeman remembered.

This accounted for the tepid reactions he got when he first pre-sented his research on penicillin at a lecture—it barely raised an eyebrow—at the Medical Research Club in London in February 1929. "He was very shy, and excessively modest, in his presentation, he gave it in a half-hearted sort of way," the chairman recalled, "shrugging his shoulders as though he were deprecating the impor-tance of what he said."

"Quiet people," Stephen Hawking once said, "have the loudest minds."

Before Fleming's discovery in 1928, a simple scratch from a rosebush could ignite a virulent infection and ear infections could spread to the brain. Strep throat, whooping cough, and scarlet fever—all rou-tinely cured with antibiotics today—killed thousands of children every year. Without effective medicine, infections spread rapidly: One-quarter of people who went into a hospital for surgery before Fleming's discovery could expect to die from an infection.

Fleming firmly believed his "mold juice" could help, even if others had doubts. "He said over and over again that penicillin had great potential therapeutic value," recalled Lewis Holt, a young scientist who worked with him. But Fleming was a microbiologist, not a trained chemist, and he did not have the resources at St. Mary's to turn his mold into a testable drug.

Knowing what we do now, it's hard not to envision a horde of researchers rushing in to figure out a way to manufacture penicillin as people were dying from untreatable illnesses. But at the time, chemists were busy studying other compounds in their lab, and none had any idea that penicillin would turn out to be such a lifesaver.

As a result, it was not until 1939 that a team of researchers led by Howard Florey and his colleague Ernst Chain at the University of Oxford took on the challenge. After managing to extract, stabilize, and purify the compound, they tested it on animals and demonstrated that it had no toxicity—a problem that had plagued earlier substances.

The drug showed clinical promise immediately. The first human to be treated, a 43-year-old Oxford policeman with a staph infection, improved rapidly after receiving an injection of penicillin in February 1941—but he died soon after, because manufacturing had not yet ramped up and not enough of the drug was available to treat him.

One year later, the compound saved the life of Anne Sheafe Miller, a 33-year-old patient at New Haven Hospital. For weeks, doctors had tried to quell a virulent bacterial strep infection Miller had contracted after suffering a miscarriage. In March 1942, with her fever spiking to 106°F, doctors managed to track down and administer a dose of penicillin—which had not yet been used in the United States—and saw her rebound overnight. Miller went on to live another six decades, becoming a grandmother and great-grandmother before dying at the age of 90 in 1999.

Fleming's midlife discovery remains one of the greatest game-changing events in modern medicine. In 1943, he was elected a Fellow of the Royal Society, London's premier scientific society, and knighted by King George VI. A year later, Fleming, Florey, and Chain shared the 1945 Nobel Prize in Physiology or Medicine for the discovery of penicillin and its ability to cure infectious diseases.

With the awards came scientific stardom, but Fleming never lost his modest touch. His admirers included Hollywood's Marlene Dietrich, who invited him to dinner while touring in London and read him his horoscope. The actress Bebe Daniels was stunned when she visited Fleming to interview him for a BBC radio show in May 1945. "I had expected to find 24 secretaries, eight guards, and I don't know what else," Daniels later recalled. Instead, she found Fleming bent over the Bunsen burner with his sleeves rolled up. He asked if she'd like a cup of tea.

Over the remaining years of his life, Fleming traveled the world, from Athens to Bombay, to visit laboratories and talk about his work. He was hailed wherever he went, but never indulged the celebrity role. One day, a couple of reporters approached him before breakfast at the Biltmore Hotel; he was staying there during a visit to Pfizer's manufacturing plant in Brooklyn, which had begun mass-producing penicillin during World War II. After the journalists asked Fleming what a great scientist like him was thinking about, he gave them a solemn look, his biographer Maurois wrote, and answered with typical dry humor: "Well, I was wondering whether I should have one egg or two."

One of Fleming's favorite pastimes was distributing medallions containing samples of his penicillin culture. He gave one of these as a gift to Dietrich and to the actress Ruth Draper after being smitten by one of her performances in London in 1946, and another to a neighbor who startled burglars away from his home. Prince Philip seemed to get one every time Fleming encountered him at a penicillin celebration, Brown notes. In 2017, a medallion owned by Fleming's niece was donated to Bonhams Auction house, where it was sold to an anonymous bidder for £11,875 (about $15,500).

∽

On the morning of March 11, 1955, Fleming began feeling ill, but he downplayed his symptoms to his second wife, Amalia Voureka, whom he had married after his 34-year-long marriage to Sareen ended with her death. He had a busy day ahead with lunch plans at the Savoy and dinner with Eleanor Roosevelt and the actor Douglas Fairbanks, Jr. When his doctor telephoned to ask if he should come over right away, Fleming told him not to rush. Not long after, he died of a massive coronary thrombosis.

Today, the antibiotic revolution that Fleming started has exploded. Every year, health care providers prescribe 270 million antibiotic prescriptions to patients in the United States. There is a price to over-reliance; the Centers for Disease Control and Prevention estimate that almost a third of those prescriptions are unnecessary and are fueling antibiotic resistance, a concern Fleming voiced as far back as 1945, when he accepted his Nobel Prize.

Fleming's legacy is also very personal, felt by individuals around the world whose lives were saved by the scientist's work. The Fleming Museum attracts hundreds of visitors every year: Schoolchildren come to learn about the scientist's breakthrough moment, medical students investigate the history of scientific research, and patients treated with penicillin pay homage to the man whose chance discovery cured them from aggressive bacterial illnesses.

Fleming's ashes are buried at St. Paul's Cathedral, near the tombs of the Duke of Wellington, Lord Nelson, and Sir Christopher Wren— an honor to the towering scientist with humble beginnings.

"I might have followed in my father's footsteps," he once said, "and have become a farmer. But fate decided otherwise."

CHAPTER 10

Eleanor Roosevelt

The Humanitarian (1884–1962)

Mid-September 1918 was a fateful time for Eleanor Roosevelt. She was with her mother-in-law, Sara Delano Roosevelt, at her upstate home in Hyde Park, New York, when a message arrived. Eleanor's husband, Franklin, then assistant secretary of the Navy, was returning from a two-month tour in Europe on the U.S.S. *Leviathan*. He was ill with the flu and pneumonia—a victim of the 1918 influenza pandemic, which had sickened passengers aboard the ship and was intensifying its lethal gallop around the globe.

Eleanor was startled by the news and anxious about her husband's health. Franklin was too incapacitated to disembark on his own, she was told, and needed immediate medical assistance. When the *Leviathan* docked in New York on September 19, four naval orderlies carried the six-feet-two-inch 36-year-old off the ship on a stretcher to a waiting ambulance, which transported him to the family's town-house on the east side of Manhattan to convalesce.

Her husband's condition was challenging enough, but the most severe shock came later, when Eleanor unpacked Franklin's luggage. There, amid his belongings, she discovered a packet of love letters from Lucy Page Mercer, a social secretary Eleanor had hired several years earlier to help her with correspondence.

Until that moment, 33-year-old Eleanor had faithfully supported her husband and cared as best she could for the couple's five children, then between the ages of two and 12. Although the letters between Lucy and Franklin have never been found—and Eleanor revealed nothing about them in her memoirs—she told her friend and biographer Joseph Lash decades later that the discovery not only stunned her, but also shattered her sense of stability. "The bottom dropped out of my own particular world," she wrote him, "and I faced myself, my surroundings, my world, honestly for the first time."

Despair tests the human soul. The discovery that her husband was having an affair with a woman she trusted came as a crushing blow. But it also ignited a fierce determination that reshaped the trajectory of Eleanor's personal and professional life. As a young child, Eleanor had withstood the deaths of her mother, father, and a younger brother—an avalanche of tragedies that forced her to scavenge and stockpile emotional fortitude. Now, as a grown woman, she went back to that reservoir and drew on it to create a purposeful new life.

Eleanor's marriage to Franklin had launched her life journey, but it would no longer dictate the road map. From that decisive day forward, she dove into public life and service—advocating for women's issues at the Democratic National Committee, arguing in favor of the International Ladies' Garment Workers' Union, delivering speeches on human rights—all along preparing herself for the role that would propel her destiny: first lady of the United States.

Eleanor Roosevelt

When Eleanor was 48 years old, Franklin Delano Roosevelt raised his hand before Chief Justice Charles Evans Hughes on March 4, 1933, and took his oath of office to become the 32nd president of the United States. It was a monumental moment for the new president—and equally so for Eleanor, who now had an international platform from which she could take a stand on political issues and advocate for human rights.

Eleanor Roosevelt's most lasting impact materialized in the latter third of her life, earning her the title of late bloomer—a designation that should be viewed not as a slight, but as an honorific. Some late bloomers experience an epiphany or a change of purpose after forging a fulfilling life in another domain. Eleanor, by contrast, evolved steadily over time. In her 30s and 40s, she began resolutely developing and communicating her views. But in her 50s and beyond, Eleanor's influence surged, and she grew into her defining legacy—one of the great humanitarians of the 20th century.

During her 12-year tenure in the White House, Roosevelt upended traditional expectations of the role of first lady by forcefully speaking out on social, racial, and religious injustice—a campaign she continued fighting long after FDR's death in 1945. At age 61, she was appointed a delegate to the United Nations General Assembly by President Truman, where she spearheaded the passage of the Universal Declaration of Human Rights in December 1948, just weeks after her 64th birthday. The document, which Nobel laureate Nadine Gordimer called "the creed of humanity," left an indelible mark on Eleanor Roosevelt's quest for justice and peace that resonates powerfully today.

A shy and fearful child, Eleanor became a public figure, public speaker, and public symbol of integrity, hope, and compassion. Her journey was inspired not by a single flash, but by multiple sparks, like a gathering of fireflies on a summer evening. First one light, then another: She reached for every one and cupped them in her hands.

In the last three decades of her life, they became a glittering orb of light—one person's offering to the greater good of the people.

"We are constantly advancing, like explorers, into the unknown, which makes life an adventure all the way," she wrote. "How interminable and dull that journey would be if it were on a straight road over a flat plain, if we could see ahead the whole distance, without surprises, without the salt of the unexpected, without challenge. I wish with all my heart that every child could be so imbued with a sense of the adventure of life that each change, each readjustment, each surprise—good or bad—that came along would be welcomed as a part of the whole enthralling experience."

On December 1, 1883, 20-year-old Anna Livingston Ludlow Hall and 23-year-old Elliott Bulloch Roosevelt united two of New York's most elite social pedigrees in marriage at Calvary Church in Manhattan. Anna, the eldest of six children, came from a wealthy and philanthropic Hudson Valley family. Elliott was a member of the Oyster Bay branch of the Roosevelt clan, and the younger brother of Theodore Roosevelt, who would later grow up to become the 26th president of the United States.

Anna and Elliott's marriage made for "one of the most brilliant weddings of the season," the *New York Times* noted the next day. The bride, celebrated for her beauty, wore a white satin dress, lace veil, and lengthy train dotted with artificial orange blossoms. She carried a bouquet of white roses and lilies of the valley; her six bridesmaids held bunches of Maréchal Niel and Jacqueminot roses. The altar was covered in tropical plants and ferns, and the pews packed with the most notable members of society, including Roosevelts, Astors, and Vanderbilts.

Well-dressed, well-mannered city hobnobbers—"swells" as the local papers liked to call them—the newlyweds set up house in a Manhattan brownstone and spent much of their time cavorting with friends at the opera, horse shows, and polo matches. Social gatherings in this gilded age were lavish and fashionable affairs. At one of these coveted events—an annual ball thrown by Mrs. William Astor in her Fifth Avenue home in January 1884—Anna and Elliott joined hundreds of other well-heeled guests decked in tulle and satin and diamond ornaments. Amid chandeliers draped in ivy, solid silver candelabra holding imported pink candles, and paintings by French artists Émile Van Marck and Jean-François Millet, they drank wine and dined on turkey, beef, duck, and candied fruits and cakes. By midnight on that day in January 1884, the gathering "was at its height," the *New York Times* reported, with guests filling the ballroom to dance. Dinner was served until three o'clock in the morning.

This was the world into which Anna Eleanor Roosevelt was born eight months later, on October 11, 1884. The first child and only daughter of the couple's three children, Anna was named for her mother, but would always be called Eleanor—a significant distinction, as it soon became clear that mother and daughter did not resemble one another either physically or by nature.

These differences were apparent from the earliest days of Eleanor's life. "My mother was one of the most beautiful women I have ever seen," she wrote in the first line of her autobiography. But that left Eleanor with a feeling of disappointment about how she herself turned out. "My mother was troubled by my lack of beauty, and I knew it as a child senses these things," Eleanor wrote. "She tried hard to bring me up well so that my manners would compensate for my looks, but her efforts only made me more keenly conscious of my shortcomings."

Looks mattered, but Eleanor's demeanor—reserved and pensive—also seemed strange to her mother. In her memoir, Eleanor remembered feeling embarrassed by her mother's gaze, as well as her tone of voice before bedtime. "Come in, Granny," her mother would say as Eleanor approached to wish her mother good night. If a visitor happened to be there, "she might turn and say, 'She is such a funny child, so old-fashioned that we always call her "Granny."' I wanted to sink through the floor in shame."

Elliott became Eleanor's refuge, and her warmest recollections of childhood center on him. Charming and athletic, filled with warmth and laughter, he adored his "Little Nell," as he used to call his daughter, after the virtuous young heroine in Dickens's novel *The Old Curiosity Shop*. Eleanor described the joy she felt as her father scooped her up and lifted her high in the air during an evening dance party in their home. "With my father I was perfectly happy," she wrote. "He was the center of my world, and all around him loved him."

But these moments were short-lived. The reality was that Elliott Roosevelt struggled with depression and alcoholism, a longtime family affliction, and spent many days and weekends away from home with his friends. Elliott's toxic binges overwhelmed his wife and horrified Theodore, who would later petition to deem his brother "insane" after watching him become increasingly irrational and even suicidal.

Much of what Eleanor remembered about her father was re-created in her imagination, notes her biographer Blanche Wiesen Cook: "She brooded over his letters, romanticized his flamboyant life, and continually enhanced and intensified memories of what were only fleeting moments." In her memoir, Eleanor readily admits that she lived in a "dreamworld," when it came to her father. He was the parent who fully accepted and embraced her—"to him, I was a miracle from Heaven" she wrote—and these recollections, no matter how imperfect, gave her comfort.

In the spring of 1887, Anna made the decision that a change of location, away from Elliott's New York bar mates, might help. On

May 18, the couple set out for Europe on the *Britannic* with 2½-year-old Eleanor in tow. What should have been a straightforward journey turned into a terrifying ordeal when the ship collided with an incoming vessel, the *Celtic,* in thick fog off the coast of New Jersey. The *New York Times* wrote about mangled bodies and passengers who heard "the piercing shrieks of the dying and the moans of the injured." The dead were dropped overboard to their briny graves.

The ordeal traumatized Eleanor, who reportedly clung to a crewmember, crying and flailing, before being dropped from the ship into her father's arms on a jolting lifeboat below. Just days after returning to land, her parents decided to restart their journey, but they left Eleanor—now terrified of water—behind in the care of Elliott's Oyster Bay aunt, Annie Gracie, who noted the effect on her great-niece: "She was so little and gentle & had made such a narrow escape out of the great ocean that it made her seem doubly helpless & pathetic to us."

It is impossible to know what ran through Eleanor's very young mind, but Annie Gracie wrote that Eleanor begged to know where her parents had gone and where she belonged. Biographers have suggested that this episode left Eleanor with feelings of abandonment, loneliness, and inadequacy. "If she had not cried, if she had not struggled, if she had not been afraid, if she had only done more and been better, she would be with her parents," wrote Cook. "This theme would be repeated again and again in her young life."

The next few years provided some happy interludes—at least from Eleanor's perspective. In the winter of 1890, when Eleanor was five, the family ventured to Italy—another effort to heal Elliott, who by then had become wildly moody and out of control. Eleanor remembered her father taking on the role of gondolier and singing with fellow boatmen as they floated down the Venice canals. "I loved his voice and, above all, I loved the way he treated me," she wrote.

But none of it lasted. In December 1892, a cascade of tragedies commenced when Eleanor's mother died from diphtheria at the age of 29, when Eleanor was just eight years old. By then, her father was living apart from his family in Abingdon, Virginia—an exile overseen by Teddy as a way to further his brother's chance at rehabilitation and also to distance the family from tension over an internal drama: Elliott's affair with Katy Mann, a servant Anna employed. Because Elliott was unable to care for his children, Eleanor and her brothers were sent to New York City to live with their maternal grandmother, Mary Livingston Ludlow Hall, in her brownstone on West 37th Street.

On Christmas that year, just a few weeks after her mother's death, her father came to visit. He filled Eleanor's stocking with special gifts—white gloves, a handkerchief, hair ribbons, and a gold pin—and gave her a puppy. But soon, he would be gone again, and heartbreak would overwhelm any hope of joy. In May 1893, Eleanor's younger brother Elliott Jr., known as Ellie, died from scarlet fever and diphtheria.

Over the two-year period while he was away, Eleanor and her father had exchanged letters, in which he expressed his love and hope for a better life together, at one point writing "maybe soon I'll come back well and strong and we will have such good times together, like we used to have." But the final calamity struck in August 1894. Grieving the loss of his wife and child and still wrestling with acute alcoholism, Elliott Roosevelt died at the age of 34. The *New York Times*'s obituary attributed the cause to heart disease, but biographical accounts suggest he suffered a convulsion after a fall.

In less than two years—and just two months before Eleanor's 10th birthday—a family of five had diminished to two: Eleanor and her three-year-old brother, Hall. What person can make sense of this? How does a child begin to recover?

Eleanor's grandmother decided not to take her grandchildren to their father's funeral. As a result, his death seemed intangible; the only way Eleanor could cope was to join him in spirit. "From that time on I knew in my mind that my father was dead," she later wrote, "and yet I lived with him more closely, probably, than I had when he was alive."

Eleanor had no choice but to survive. Life with Grandmother Hall was structured and strict—she believed her grandchildren needed the discipline her own children lacked. Eleanor remembered dressing in warm flannel clothing from fall through the beginning of spring. Summer was a welcome respite. Eleanor cherished visits with her Oyster Bay side of the family at Sagamore Hill, where her uncle Ted chased the kids past haystacks, read poetry aloud, and enveloped his niece in affection. At her grandmother's country house in Tivoli, a village in New York's Hudson Valley, Eleanor rode her pony named Captain, played tennis, and rowed boats with her aunts and uncles. She grew especially close to two of them: Pussie, who read poetry with her, and Maude, whose kindness comforted her. There were picnics and campfires—and she relished the solitude of the country, often reading a book in the fields or under a tree.

During these early years, Eleanor received required instruction for a young girl—how to wash, iron, and sew—and she spent hours walking with a stick behind her shoulders to improve her posture. But what would she *become?* She was reticent and tall (she would soon grow to her full five feet 11 inches) and awkward. What *could* she become?

She wanted to be a singer, Eleanor later wrote. Her father had encouraged her to be musical, and she loved to listen to her aunt Pussie play the piano. But it was more than just the music that appealed to her; "Attention and admiration were the things through all my child-hood which I wanted," she wrote, "because I was made to feel so conscious of the fact that nothing about me would attract attention or would bring me admiration!"

Spark

⌒

Eleanor's first inkling of independence came at age 15, when she was sent abroad to the Allenswood boarding school just outside central London. Here, beginning in the autumn of 1899, she would be educated under the wise mentorship of Mademoiselle Marie Souvestre, the school's director, who had previously taught another Roosevelt: Eleanor's Auntie Bye, her father's older sister.

Over the next three years, Eleanor studied most of her core subjects in French, which she had learned at school in New York, and also took German and Latin. Allenswood inspired a sense of confidence and self-reliance that she had been lacking. She played field hockey and took trips abroad to St. Moritz, Florence, and Paris, where she ordered a dark red dress from a French clothier—a far cry from the conservative clothes on which her grandmother Hall had insisted.

Marie Souvestre not only oversaw Eleanor's education, but also reared her into maturity. In her young charge, she saw a lively mind and an ability to think clearly—qualities she admired and sought to nurture. Short and stout with snowy hair, she encouraged Eleanor to experiment with fresh ideas, rather than parrot back what she had learned. Eleanor especially enjoyed the evenings when Mademoiselle Souvestre invited students into her study, where she read poems, plays, and stories aloud in French.

During the final few months of her schooling, the two traveled together on holidays to Calais, Frankfurt, and Rome, where Mademoiselle Souvestre brought history to life as she talked about Julius Caesar's assassination on the ides of March in 44 B.C. Her impact on Eleanor's life cannot be overstated, Eleanor's grandson David Roosevelt tells me one day as we chat on the phone. His grandmother kept a photo of Souvestre "in her presence practically at all times," says David, whose father, Elliott, was Eleanor and Franklin's fourth child. "If there was ever a mentor, this was Eleanor Roosevelt's mentor."

Eleanor had no desire to leave the intellectual and independent life she so loved. But as her 18th birthday approached in 1902, her grandmother called her back to New York to make her social debut. Unlike Eleanor's mother, Anna, who had delighted in these affairs, Eleanor dreaded the notion of dressing up for parties. She didn't dance well, and she knew barely anyone. "I do not think I quite realized beforehand what utter agony it was going to be," she later wrote, "or I would never have had the courage to go."

Although she was required to endure numerous debutante dances—she was, after all, the niece of then president Theodore Roosevelt, who took office in 1901—Eleanor refused to make socializing her full-time occupation. Volunteer work ran in the Roosevelt bloodline, and Eleanor had been schooled in it at an early age by her parents and grandparents. Her studies at Allenswood and trips to Europe illuminated social disparities, and now, in New York City, she sought opportunities in philanthropy and charity.

In a prelude to what would become Eleanor's lifelong commitment to serving the underprivileged and advocating for human rights, she volunteered for the Consumers League, scrutinizing working conditions in garment factories and department stores. And she joined the Junior League, a group of young women intent on social reform, making her way through the streets of the Lower East Side to teach dance and calisthenics to children at a settlement house.

One summer day, while riding a train from Manhattan to Tivoli, Eleanor bumped into Franklin Delano Roosevelt, a Harvard junior and a distant cousin. (Eleanor's grandfather, Theodore Roosevelt, Sr., and Franklin's father, James Roosevelt, were fourth cousins.) The two had seen each other on prior occasions—even dancing at a family Christmas party when they were young teenagers—but this meeting and others that followed prompted a connection that lasted. They talked for hours, wrote letters once Franklin was back at Harvard, and visited on holidays.

The two were opposites in many ways: FDR was outgoing, sociable, and confident around girls whereas Eleanor was quiet, modest, and romantically inexperienced. It is difficult to pinpoint what attracted them to each other; Eleanor later burned Franklin's letters to her, saying they were too private. Even son James Roosevelt later wrote that "anyone who pretends to know is not telling the truth."

Still, Eleanor and Franklin were intellectually matched and civically driven. From the start, Eleanor exhibited a loyal dedication she would tender throughout their relationship. In a letter to a cousin, she wrote: "I can only say that my one great wish is always to prove worthy of him."

In Eleanor's view, it was time to move forward into adulthood. When her grandmother asked if she was sure she was in love, Eleanor said yes, despite later admitting that she had little concept of what love really was. "There seemed to me to be a necessity for hurry," she wrote, "without rhyme or reason I felt the urge to be a part of the stream of life."

Franklin, 23 years old, and Eleanor, 20, were married on the afternoon of Saint Patrick's Day, 1905. Only days earlier, Uncle Ted, with whom Eleanor had always had a close and loving relationship, had been inaugurated for a second term as president. He walked his niece, carrying a bouquet of lilies of the valley, down the aisle, and then stole the show at the couple's intimate reception, regaling their guests with stories.

Married life was not quite the stream of life Eleanor had envisioned. Her newfound independence would soon be squashed by the conventional demands of marriage and the restrictions of a willful and domineering mother-in-law. Sara Delano Roosevelt had an exceptionally strong bond with her only child and thrust herself into the couple's life, leaving Eleanor with a feeling of inadequacy. After Sara purchased and decorated the couple's first home on East 65th Street, Eleanor broke down in tears, unhappy to live in a house that did not feel like her own.

Motherhood added a different kind of challenge. Between 1906 and 1916, as Franklin advanced in his political career—first as a New York state senator in 1911 and soon after as assistant secretary of the Navy—Eleanor delivered one girl and five boys. After her third child, Franklin D. Roosevelt, Jr., succumbed to the flu at just seven months old, Eleanor blamed herself, thinking she had left him too much with his nurse. But she had little time to grieve. "For ten years I was always just getting over having a baby or about to have one," she wrote.

The most significant crisis by far—the event that would redefine the trajectory of her life—was the discovery of Lucy Mercer's letters to Franklin. Lucy was 22 years old when Eleanor hired her as a social secretary in 1913. Charming, well mannered, and reliable, Lucy handled her job efficiently and came to be accepted as a family companion—so much so that when Eleanor was short a woman for lunch or dinner, she invited Lucy. Although she was clearly aware of her husband's flirtatious habits, her grandson David tells me, she likely never dreamed he was capable of an affair: "I don't think my grandmother ever took that seriously until those letters."

In childhood, Eleanor's parents had abandoned her by dying; now, her husband had deserted her emotionally. "She felt alone, as she had before," David's wife, Manuela Roosevelt, chair of the Eleanor Roosevelt Val-Kill Partnership, tells me. And yet, in times of crisis and loneliness, Eleanor managed to dig deep and push forward. She was thoughtful and considered. "She had a very powerful inner life," says Manuela, and there, she found strength. "She always rose up."

After discovering the letters, an emboldened Eleanor rejected her selfless dedication to husband and mother-in-law and sought to define her own interests and her own mission in life. Divorce was out of the question. Eleanor offered Franklin "his freedom," as biographers describe it, but the dissolution of their marriage would have likely sabotaged Franklin's political career. Above all, Sara wouldn't allow it, even threatening to cut off Franklin's income if he didn't put an end to seeing Lucy. He agreed and Lucy went on to marry a wealthy widower, Winthrop Rutherfurd.

In the years following, Eleanor and Franklin began cultivating the relationship that would sustain them throughout their 40-year marriage: a fiercely committed partnership. They spent quite a bit of time apart, both emotionally and physically. Eleanor had never been overly demonstrative toward her husband, and she established distance in the aftermath of finding out about his affair, her son James wrote. At times, James saw his father attempt to embrace his mother, but she refused. Still, there was love between them, says David. "That's the only way their partnership could have been developed." Eleanor created her own life alongside her husband's political endeavors and dedicated herself to him in a new way.

Two other pivotal moments bolstered Eleanor's resolve to carve out her own identity. In August 1919, Grandmother Hall died at the age of 76 in Tivoli. Mother to seven children of her own and surrogate mother to Eleanor and Hall, she had spent her life focused on family—a fact not lost on Eleanor, who wondered what might have happened to her grandmother if she'd had the opportunity to pursue her own friends and interests, including an early talent for painting.

"If she had some kind of life of her own, what would have been the result?" Eleanor asked. "Would she have been happier, and would her children have been better off?"

Instead, the life her grandmother led, as much by necessity as choice, highlighted the consequences of abandoning one's self. At an early age, Eleanor vowed that she would never be fully dependent on her children; her grandmother's death only reinforced this determination. "She had to find other outlets for herself," says David, "and that's what she did."

The second consequential event happened two years later, when Franklin was diagnosed with polio after falling off his yacht into the frigid waters of the Bay of Fundy near the Roosevelt's summer home on Campobello Island in New Brunswick, Canada. Although Eleanor agreed to let FDR return to his mother's estate in Hyde Park to recuperate, she refused to yield to Sara's opinion that Franklin should put his political aspirations aside for good and settle into a quiet life for the long term.

After consultation with Franklin's physician, and with the support of her husband's confidant and political adviser, Louis Howe, it was decided that work would ultimately be the best remedy for FDR's recovery. Naturally optimistic and determined to walk—even if it meant heavy braces—Franklin mustered the stamina to persist in his political aspirations. Eleanor later reflected that his condition "gave him strength and courage he had not had before" and this was not something his mother could take away. Sara had dominated their life for years, Eleanor later reflected, and now, at age 36, she had the power to resist: "His illness finally made me stand on my own feet in regard to my husband's life, my own life, and my children's training."

Over the next decade, as a nurse, governess, and servants looked after their children and home, Eleanor dove into her own political causes. She became active in the New York State Democratic Committee, campaigning for Alfred Smith in his successful bid for governor against her own cousin, Republican Theodore Roosevelt, Jr. She advocated for women's rights through the League of Women Voters and picketed in support of shorter working hours with members of the Women's Trade Union League. And she helped oversee a public competition to come up with a plan for the United States to work with other countries to preserve world peace.

Sara did not like the idea of women in politics nor did she like Louis Howe, who wore wrinkled clothes and smelled like cigarettes—not exactly the highbred society type the Roosevelt family was used to. But determined to free herself from Sara's clutches, Eleanor allowed Howe to become her guide and mentor. He soon became the backbone of Franklin and Eleanor's political alliance, and their individual political successes.

At Howe's request, Eleanor reviewed Franklin's speeches. She also accepted Howe's tutoring in public speaking, which included learning how to control her high-pitched voice. Howe was able to penetrate her facade, son James wrote: "It was then that she first started to emerge from her shell, that she first started to realize her hidden potential."

Despite her reticence, Eleanor learned to accept the demands of public life. She learned new skills and found purpose for traits she had developed early in life. "The interest, the curiosity, fostered and nurtured in childhood were there waiting for the proper moment of expression," she wrote in an article for *Woman's Home Companion* in 1932. Now was the time to use them.

In midlife, Eleanor Roosevelt transformed herself with a simple mission no matter how daunting: Be useful, make a difference, *do* something. "Courage is more exhilarating than fear and in the long run it is easier," she wrote. "We do not have to become heroes overnight. Just a step at a time, meeting each new thing that comes up, seeing it is not as dreadful as it appeared, discovering we have the strength to stare it down."

⌒

Eleanor Roosevelt's resilience made a lasting mark on her children and grandchildren. In his memoir, James Roosevelt described the picture that best captured his mother's spirit: "striding across a deserted train station, suitcase in hand, on the move, so briskly that a younger person could not keep up, carrying her own luggage, pulling her own weight, a sort of lonely warrior fighting for those she believed in."

Grandson David marveled at the ability of his Grandmère, as her grandchildren called her, to overcome personal defeat and tragedy and "emerge stronger, more knowledgeable, more committed." She never gave up, he tells me. "Whatever the challenge was, she felt you had to face it. You may not overcome it, but you had to try."

How did Eleanor Roosevelt find the strength to bear personal calamities and move on? Where did her resilience, defined loosely as the ability to function normally despite high levels of stress and trauma, come from? Where does anyone's?

This is a question that has captivated and challenged developmental psychologists for a very long time. There's no easy way to define resilience, let alone test for it. But a researcher named Emmy Werner made significant headway when she published a landmark report in 1989 in *Scientific American* that examined how children respond over time to early stressors in their lives.

Then a professor and child psychologist at the University of California, Davis, Werner and her colleagues followed a group of 698 children born in 1955 in Kauai, Hawaii, from before birth through their third decade of life. Scientists recorded the child's family environment and stressful events that occurred. The children were assessed at age one, two, 10, 18, and either 31 or 32 to see how these many medical and environmental factors affected their ability to cope.

The majority of the children were lucky enough to grow up in stable middle-class homes, which provided clear advantages: Their parents had finished high school, and they were nurtured and supported. But a third of the cohort was not so fortunate. They endured difficult challenges—including complications first during their mother's

Spark

pregnancy, labor, or delivery, then chronic poverty, family discord, parental alcoholism, and mental illness. And they suffered as a result. By their teen years, many had developed serious learning problems or mental health challenges; others had run-ins with the law.

Remarkably, however, a third of the children in this most vulnerable group grew up to become not only functional adults, but successful in school, at home, and in their social lives. They were resilient, despite the odds, the researchers concluded, and became "caring people who expressed a strong desire to take advantage of whatever opportunity came their way to improve themselves."

The researchers identified several characteristics that seemed to gel in these children. Certain aspects appeared to be constitutional: The children were likeable and easygoing and had no eating or sleeping problems that caused their parents distress. They also established close bonds with at least one caregiver and were able to solicit support from others if a biological parent was unavailable. The children found emotional sustenance outside their immediate families, often at school, where a favorite teacher became a role model. Interestingly, girls tended to fare better than boys overall, and girls who took care of younger siblings seemed to develop a sense of independence and responsibility that shaped them as adults.

It's hard not to hear intriguing echoes of Eleanor Roosevelt's experience within these findings. Although she came from a privileged background, her childhood was infused with the strain of her father's alcoholism. She found support in several adults in her early life, especially her uncle Ted and her teacher Mademoiselle Souvestre. Before her father died, he had stressed to Eleanor that she must look after her younger brother Hall, and she fully accepted this responsibility. She accompanied him to boarding school at Groton, the Roosevelt family tradition, and wrote him daily letters while he was there. It was Eleanor whom Hall consulted when he sought to dissolve his marriage, and Eleanor who supported him during the final years of his life. Like their father, he struggled with alcoholism and died at the age of 50.

Resilience is not a personality trait, researchers believe, but rather a dynamic process that builds over time and leads to a person's ability to adapt favorably to life's curves. In a 2005 follow-up report to her original study, Werner stressed that resilient individuals were not "passively reacting" to difficult events. "Instead, they actively sought out the people and opportunities that led to a positive turnaround in their lives."

Today, scientists are bearing down at the neural level in a developing field known as the neuroscience of resilience. Their goal is to identify biological factors that may contribute to how resilience is expressed, including genes and hormones like cortisol, which drives the human response to stress. Brain scans, meanwhile, are beginning to illuminate fingerprints of resilience in the brain. Already, scientists have found that maltreatment and abuse in childhood may change the brain's architecture. One report found that severe neglect shrinks the hippocampus, an outcome that has been linked to depression and other psychiatric illnesses.

Efforts to spotlight the components of resilience might one day lead to new approaches to promoting it in those who need it most.

⌐⌐

Eleanor Roosevelt was 48 years old when FDR was elected to his first term as president. This was not a celebratory moment—at least not for Eleanor. She recognized FDR's natural gifts as a politician: his intelligence, charisma, and dedication. And she applauded his desire to serve his country and his confidence in doing so at a time of domestic and international turmoil.

But Eleanor had no interest in becoming first lady, which she worried would confine her to the White House and restrict her independence. "As I saw it, this meant the end of any personal life of my own," she wrote. She went to bed the night of the election, she later recalled, with "turmoil in my heart and mind."

The reality, of course, was that she had no alternative. Rather than reject her new position, she dramatically redefined it. First ladies had long acted as close advisers to their husbands and engaged in volunteer work, often around family welfare and education. But none had stepped in with the tenacity of Eleanor, who seized the role with a megaphone and became an activist for the causes she believed in.

Eager to spend time outside of the White House, Eleanor sought moments of opportunity. In April 1933, just weeks after FDR's inauguration, the first lady broke up a dinner party when guest Amelia Earhart invited her to fly round-trip from Hoover Field in Arlington to Baltimore. FDR was away, but Eleanor and the rest of the party, which included her brother Hall, hopped aboard Earhart's twin-engine Curtiss Condor plane. Earhart still wore her white evening dress and Mrs. Roosevelt sat nearby in her blue dinner dress topped by a straw hat. With city lights sparkling below, the first lady told reporters she felt "on top of the world," and fully trusted her female pilot. "I'd give a lot to do it myself," she said.

In prior political missions with her husband, Franklin had taught Eleanor to become a careful reporter, observing every detail—from soil erosion to wash hanging on the line—as a way of better understanding the conditions affecting the land and its people. Eleanor relied on this skill as she famously became Franklin's "eyes and ears," reporting back vital information to the White House and relieving him from the exhaustion of traveling with heavy leg braces.

Nothing could stop her. During FDR's first term, she visited destitute areas in Puerto Rico, where typhoid was common; she championed construction of a controversial housing community for impoverished laborers in Arthurdale, West Virginia; and she investigated working conditions for coal miners. In Bellaire, Ohio, she donned a miner's cap and traveled deep into a two-mile stretch underground where she spent close to two hours watching 400 workers. Although a telephone hookup was provided so she could talk to the president directly from the mine, she said it would be better not to bother him. Later, after emerging from underground, thousands of citizens greeted her on flag-lined streets in the nearby town of Steubenville.

Despite the Secret Service's best efforts to accompany her wherever she went, the first lady insisted on driving her own car, especially if it meant she could sneak out of the White House for lunch or tea with friends, like Democratic activist Elinor Morgenthau. She refused to have a security agent shadowing her, but agreed to carry a pistol and practiced shooting at targets. "I would never have used it on a human being," she wrote, "but I thought I ought to know how to handle a revolver if I had to have one in my possession."

Over the course of FDR's 12 years in the White House, Eleanor's passion for civil rights burgeoned into a cause she would fight for the rest of her life. In 1934, she joined the National Association for the Advancement of Colored People (NAACP) and doggedly supported anti-lynching legislation, even though FDR refused to back it, fearing he would lose the support of white Democratic voters in the South. "She was his conscience," wrote Cook, "and she knew it." According to biographer Lash, at a meeting with NAACP leader Walter White at the White House, FDR turned to White at one point and said, "Somebody's been priming you. Was it my wife?"

One of the first lady's most public stands took place in 1939, when she wrote a letter of resignation to the Daughters of the American Revolution (DAR) after the organization barred Marian Anderson, the African-American contralto singer, from performing at Constitution Hall. At Mrs. Roosevelt's invitation, the singer had performed at the White House several years earlier, and she could not allow the

DAR's ban on Black entertainers to stand without serious reproach. "I am in complete disagreement with the attitude taken in refusing Constitution Hall to a great artist," she wrote. "You had an opportunity to lead in an enlightened way and it seems to me that your organization has failed."

A poll showed that two-thirds of Americans approved of Mrs. Roosevelt's resignation, and weeks later, on Easter Sunday, Anderson performed instead on a stage built over the steps of the Lincoln Memorial. Wearing a fur coat, the singer began her concert with "My Country, 'Tis of Thee" before a crowd of 75,000 people. "Genius draws no color line," said Secretary of the Interior Harold Ickes, who introduced her, "and so it is fitting that Marian Anderson should raise her voice in tribute to the noble Lincoln, whom mankind will ever honor."

The traumas in Eleanor's life had imbued her with a vast reserve of empathy; her genius was in how she offered it to others—in a straightforward, non-sanctimonious way. Communication turned out to be the first lady's forte. Just two days after FDR's inauguration, she appeared at the first of 348 press conferences she would hold for female journalists, elevating their status as reporters and ensuring some their jobs. Every one was a respectful "battle of wits," she later recalled, allowing her a voice separate from the president. "As time went by, I found that people no longer considered me a mouthpiece for my husband, but realized that I had a point of view of my own with which he might not at all agree."

In an extraordinary article, "In Defense of Curiosity," published in the *Saturday Evening Post* in 1935, a 50-year-old Eleanor explained why the pairing of intellectual and emotional curiosity was so critical to a fulfilling life, including her own. "I was born with an interest in people," she wrote. And although life would always be filled with sorrow, "curiosity will insure an ever-recurring interest in life and will give you the needed impetus to turn your most baleful experience to some kind of good service. A little of the right kind of curiosity will endow you with sympathy, and that will bring you the confidence of your fellow human beings."

Curiosity inspired knowledge, thoughts, and opinions, which Eleanor Roosevelt shared directly with the public in radio shows, speeches, and books. "My Day," her daily newspaper column, was like a modern-day blog—documenting everything from FDR's New Deal activities in the 1930s and her outspokenness over the plight of Jewish refugees in the '40s to her takedown of Senator Joe McCarthy's "Red Scare" in the '50s.

Politics was not her only focus. Mrs. Roosevelt's ability to connect with ordinary Americans attracted a following of eager groupies who sought her advice. In the first lady's popular "If You Ask Me" column, which ran from 1941 to 1962, she answered any question her readers posed. "She saw them as individuals," her granddaughter Nancy Roosevelt Ireland wrote, "and many of them saw her as a friend."

Eleanor's honesty appealed to her supporters, and she never minded being blunt. When asked if she thought 67 is too old for psychoanalysis, she wrote, "I don't suppose one is ever too old for anything. But if at 67 you have not learned to know yourself, it is a little hard for me to believe that anyone else can help you do so." When one reader called her out for pinning a corsage on her left shoulder—"Isn't a corsage supposed to be worn on the right shoulder?" she asked—Eleanor responded without a hint of apology: "I haven't the faintest idea. I thought corsages were worn where they felt most comfortable and looked best. I didn't know there was any rule, and I am afraid I don't care!"

Through it all, the first lady put up with jeers, personal attacks, and even death threats. Cartoons featured her jutting front teeth; she was mocked for the altitude of her voice, for traveling too much, and for spouting too many opinions. In 1958, when she was 73 years old, the Ku Klux Klan in Monteagle, Tennessee, put a $25,000 bounty on her head when they learned she was coming to town to conduct a workshop on nonviolent disobedience for civil rights. The FBI followed her for two decades, creating a 3,000-page file between 1940 and 1961, "due to her prominent public role."

She found her greatest support among her children, and later her grandchildren. She also depended on a circle of women who became her closest confidantes, including suffragists and political organizers Nancy Cook and Marion Dickerman, and Associated Press reporter Lorena Hickok, with whom she exchanged amorous letters, leading some to speculate the two had an intimate relationship.

But she also found friendship in strangers, the people she met on her travels throughout the country and the world. Every person has surprising and unexpected qualities, she wrote. "The treasure is there if you will mine for it."

On April 12, 1945, Franklin Delano Roosevelt died after suffering a cerebral hemorrhage while sitting for a portrait in his cottage in Warm Springs, Georgia, a place he visited regularly to take in the therapeutic waters after his polio diagnosis. He was 63 years old. Upon learning the news, Mrs. Roosevelt was reported to have said to the president's doctor and secretary: "I am more sorry for the people of the country and the world than I am for us." Eleanor later disputed that she made the statement, but it made newspaper headlines and carried a sentiment of generosity admired in the first lady. In a telegram to her children, she wrote: "Darlings: Father slept away this afternoon. He did his job to the end as he would want you to do. Bless You. All our love. Mother."

The president's death reignited old anguish when Eleanor discovered that Lucy Mercer Rutherfurd had been visiting Franklin when he died and that the two had met at other times during his presidency as well. It also raised questions about what would become of Eleanor, now that her husband was gone. Despite her many accomplishments as first lady, she felt she had lived her years in the White House in a dispassionate way. "It was almost as though I had erected someone outside myself who was the President's wife," she wrote in her memoir. "I was lost somewhere deep inside myself."

She would find herself where she always did, at Val-Kill, the rustic cottage built especially for her two miles from the Roosevelt family house, "Springwood," in Hyde Park. Eleanor treasured Val-Kill, the only place that was truly hers, and the land it inhabited next to a gurgling stream. Here she could spend the night on the sleeping porch, unload the weight of her public persona, and recharge her spirit. And here she retreated after FDR's death.

On a recent spring visit to Val-Kill, where daffodils splashed yellow all around, I was struck by the plainness of Eleanor's cottage with its wood-paneled walls and cozy unmatched armchairs. It felt like a family home—I recognized her apple dish pattern as one that belonged to a friend—filled with the warmth of family and friends whose photographs covered the walls. The natural simplicity was a far cry from the opulent homes her mother had circulated through at social parties in New York City.

Walking through the woods outside, I found "Eleanor's Walk," a one-mile trail where trees filter light and winter leaves crunch beneath your feet as spring begins to awaken. I relished in the quiet and the meditative feeling that Eleanor experienced herself decades back as she walked through the woods with her grandchildren and listened to the sounds of nature. "Those were very special times," says her grandson David, who lives with Manuela and their children in a house on the Roosevelt estate. "She would encourage us to open up, to talk to her, to ask her questions, to tell her what was going on in our lives."

The camplike setting provided the ideal place for family to gather. Eleanor took dips in the pool and counted red-winged blackbirds over breakfast. David would often see his grandmother propped up against a tree with a book. "As small children, we knew, Grandmère is out reading so we don't want to bother her. We sensed that it was her time to be alone," he tells me. "And yet, I have to think now, how often as she was sitting under the tree was she really reading? Or was she contemplating much higher issues?" This peaceful refuge, her grandson says, provided his grandmother with a place to not only rejuvenate, but also to develop her thinking on social and political issues.

At Val-Kill, Eleanor entertained a wide range of political figures, both before and after her husband's death. John F. Kennedy met with her in the nook of her living room to seek support for the presidency. Emperor Haile Selassie of Ethiopia, and India's Prime Minister Pandit Jawaharlal Nehru also visited. She served hot dogs to King George VI and Queen Elizabeth.

One day, 10-year-old Shirley Temple arrived wearing a starched dress and white gloves and carrying a lace purse that held her beloved slingshot. The two walked through the grounds and later, as Mrs. Roosevelt tended to chops on the barbeque, Shirley picked up a pebble, placed it in her slingshot, pulled back the elastic bands, and took aim, hitting the first lady square in the backside. Mrs. Roosevelt, Shirley noted, jerked and thrust her barbecue fork in the air "like the Statue of Liberty," but never looked back to identify the troublemaker. Shirley's mother gave her a slap on the bottom.

After FDR's death, a 60-year-old Mrs. Roosevelt famously told reporters, "The story is over." But there was so much more to come.

Over the next 17 years of her life, Eleanor Roosevelt, the late bloomer, became a truly independent activist, writer, and stateswoman—free from the spotlight of her husband's presidency and eager to make change.

In 1946, President Truman appointed her a United States Representative to the United Nations, a position she accepted, but not without "fear and trembling" over whether she was qualified. She served until 1953.

In one of her proudest campaigns, she became the driving force behind the adoption of the Universal Declaration of Human Rights, which outlined fundamental rights for all, including education, liberty, and security. The declaration, since translated into more than 500 languages, from Arabic to Zulu, was adopted by the United Nations in 1948, when Eleanor was 64 years old.

In the years following, she spoke out on a slew of issues, including school integration and workers' rights. In 1959, as a member of the National Advisory Committee on Farm Labor, she testified before Congress in support of a minimum wage and for unemployment and workmen's compensation laws. Reflecting back to her teenage years in Manhattan, where she surveyed the city's sweatshops, Eleanor recalled images of women and children toiling in crowded dirty workplaces to earn a few pennies. "These conditions I can never forget," she testified. In her 70s, she campaigned for Kennedy's presidential bid and, at his request, chaired the President's Commission on the Status of Women.

Mrs. Roosevelt seemed incapable of stopping. She had to help and do and teach. At the age of 75, she took a new job as a lecturer at Brandeis University. "I suppose I should slow down, but I could not, at any age, be content to take my place in a corner by the fireside and simply look on," she said. "One must never, for whatever reason, turn his back on life."

Spark

By the end of her life, Eleanor Roosevelt had 22 grandchildren and 13 great grandchildren. None of their antics fazed her—not even when David, then about six or seven, ran from the pool to the restroom in the Val-Kill cottage. "I was wet and dripping water all over," says David. "As the story goes, my grandmother turned to Madame Chiang Kai Shek and said, 'That was my grandson. I'll introduce you later.'"

Eleanor's escapades behind the wheel didn't faze her family much, either, even though she was a notoriously bad driver. "Wherever we went, people recognized her Studebaker car and they would make room for her," David says with a laugh. "And it wasn't out of respect. It was out of self-protection."

Today, many visitors plan a pilgrimage to Val-Kill to see the place where Eleanor Roosevelt spent the final years of her life. Springwood gets more visitors, supervisory park ranger Franceska Macsali-Urbin tells me, but Val-Kill is most important to those who considered the first lady an example for humility, good deeds, and human rights. Her unwillingness to cave to criticism and her determination to help—as unpretentious as it was—awed those who wondered if they could ever muster her degree of courage. "People get emotional, they cry that they're here," says Urbin. "I don't know how many times women have said, 'I admire her so much, she's my hero.'"

It is unlikely that Eleanor Roosevelt would have been comfortable with that description. She felt she had an obligation and a duty to serve. "She just did what she had to do," David tells me. "There was no resting on her laurels."

In the spring of 1960, Eleanor Roosevelt was diagnosed with aplastic anemia, a condition in which the bone marrow fails to make an adequate number of blood cells. Saying she was "too busy to be sick," she continued to travel, but her condition worsened the next year. In the summer and fall of 1962, Eleanor endured numerous hospitalizations and procedures, including blood transfusions and painful bone marrow aspirations. After a series of inconclusive tests, doctors finally diagnosed her with tuberculosis in October—days after she had demanded to be released from the hospital so she could spend her last days at home. On November 7, 1962, Eleanor died at the age of 78. The first lady had requested a modest funeral, but Presidents Truman, Kennedy, and Eisenhower and Vice President Johnson all showed up.

"In the White House and for some time thereafter," the *New York Times* wrote in its obituary, "no First Lady could touch Mrs. Roosevelt for causes espoused, opinions expressed, distances spanned, people spoken to, words printed, precedents shattered, honors conferred, degrees garnered." The unlikely leader had become a global humanitarian.

Most remarkably, she did it all in the second half of her life—an independent late-blooming woman with the determination and energy of a cavalry.

"People develop at different ages and overcome things at different times, so you cannot expect that everyone will feel that the most rewarding years of their lives came during the time of youth," she once wrote. "It may well be that one's development came later and therefore one will feel the results at a later period."

CHAPTER 11

Peter Mark Roget

The Wordsmith (1779–1869)

I n May 1799, one year after graduating from medical school, Dr. Peter Mark Roget found his way to the Pneumatic Institution, located in Clifton, England. The new establishment, which was designed to study the medical potential of various kinds of inhaled gases, was run by a portly man named Thomas Beddoes.

Eager to cure lung ailments, including tuberculosis, Beddoes and his lab scientist Humphry Davy were investigating nitrous oxide, one of a number of gases being studied for its therapeutic effects. Before giving it to sick patients, however, Davy decided to test the substance on himself and volunteers he recruited to aid in his research.

Twenty-year-old Roget joined a star-studded lineup of society's elites who came to Davy's aid, including Thomas Wedgwood, heir to the Wedgwood pottery empire, and the poets Samuel Taylor Coleridge and Robert Southey. After inhaling the gas through a silk bag, participants quickly discovered its mind-altering impact. Laughing gas, as it was later nicknamed, prompted fits of involuntary giggles and intoxicating euphoria.

In accounts Davy later published, volunteers described feeling cheerful, powerful, and never more alive. One participant said he began leaping around the lab as if an actor on stage; a music lover said the intensity of delight compared favorably to listening to the *Messiah* sung by a chorus of 700 at Westminster Abbey. Coleridge experienced warmth, as if he had just returned to a toasty room after a walk in the snow, and described "more unmingled pleasure than I had ever before experienced." Wedgwood felt so light he thought he might float to the ceiling. Southey reported a tingling in his fingers and toes and even a "thrill in my teeth."

Amid all this glee, Peter Mark Roget had a reaction that can best be described as unenthusiastic. In a somber account, Roget chronicled a heavy fatigue that morphed into delirium, heart palpitations, vertigo, trouble breathing, agitation, confusion, and the inability to speak. The gas completely upended his equilibrium, he reported, even his ability to think. "Thoughts rushed like a torrent through my mind, as if their velocity had been suddenly accelerated by the bursting of a barrier which had before retained them in their natural and equable course," Roget wrote. "I cannot remember that I experienced the least pleasure from any of these sensations."

Roget's experience with nitrous oxide was brief and relatively insignificant in the scheme of his life journey—but it provides a perfect metaphor for a man whose overarching mission was to make order out of chaos. Unlike his compatriot volunteers, who were astonished by the effects and took delight in losing control over their senses, Roget felt lost, disoriented, and above all, disorganized.

These were feelings he struggled to remedy throughout his life, beginning in early childhood, when he experienced the death of his father and the nomadic lifestyle his mother instituted. Roget discovered that words helped him bear the dysfunction at home, and he became transfixed by the power they offered: If he could not keep his family in check, at least he could systematize vocabulary in distinct categories in a notebook.

Roget's childhood passion never left him. In 1805, when he was 26 years old and beginning to practice medicine, he drafted a list of some 15,000 words organized into tidy groupings. But it was not until 44 years later, when Peter Mark Roget was 70 years old and newly retired from his career in science and medicine, that he resurrected his work and hammered out the vast compendium of words that would define his legacy: *Thesaurus of English Words and Phrases, Classified and Arranged so as to Facilitate the Expression of Ideas and Assist in Literary Composition,* or, as it is widely known, *Roget's Thesaurus.*

Some late bloomers come to a newly discovered avocation after a lifetime in an unrelated field; others, like Roget, rediscover an interest that surfaced in childhood. Roget's passion for words bookended his life, emerging in his youth and resurfacing late in life—not unlike a favorite tie that hangs in the back of a closet until the moment is right to wear it.

Peter Mark Roget, the polymath, lived a long and prolific life. Over the course of his professional career, he practiced medicine, gave scientific lectures, reported a breakthrough observation in optics that would later inform cinematography, wrote a treatise on animal and vegetable physiology, and even published a paper on how to move a knight effectively across a chessboard.

But Roget's late-blooming passion earned him perpetual renown and gifted his love of language to future generations. Barbara Ann Kipfer, editor of the eighth edition of *Roget's International Thesaurus,* published in 2019, says she finds it thrilling to open his book and find herself awash in words. Even more impressive is the depth and

Spark

scope of his work. "It is heavenly," she tells me. "I absolutely cannot conceive of how he did it."

⸺

Peter Mark Roget was born into a family filled with happy potential—if only it had lasted. His father, Jean Roget, a transplant from Geneva and pastor at Le Quarré, a French Protestant church in Soho, was warm, animated, and smitten with Catherine Romilly, a Brit who loved French literature. Although Jean had no fortune to offer his bride, the well-off Romillys liked him and readily accepted him into the family.

The couple married in February 1778. Peter Mark Roget—named for his grandfathers Peter Romilly, a British jeweler, and Jean Marc Roget, a Swiss clockmaker—was born one year later, on January 18, 1779, in London's immigrant-rich Soho neighborhood. The baby's arrival thrilled his parents. "The happiness they enjoyed upon their marriage was as pure, and as complete, as is ever the portion of human beings," Catherine's brother Samuel later recalled, "but it was of very short duration."

Just a few months after Peter's birth, Jean became ill with tuberculosis. Following the advice of doctors, Catherine took her husband back to his native Geneva, leaving their son with her father, who looked after the baby with help from a nursemaid. Catherine's mother played little role in her grandson's life. She had lost six of her nine children—a torrent of tragedies that contributed to what appeared to be a lifelong battle with depression.

Jean Roget's health did not improve in Geneva. When Peter was two years old, the Rogets decided that if their son had any hope of knowing his father, he needed to join them abroad. Peter's doting grandfather reluctantly agreed to let the boy go. Peter's uncle Samuel, who had become a close friend of Jean's, transported his nephew across the English Channel to be reunited with his parents.

Just after the birth of the couple's second child, Annette, in April 1783, Jean succumbed to his illness. He was in his early 30s, leaving an even younger Catherine a widow with two small children. In a letter to his sister after Jean's death, Samuel encouraged Catherine to think about the blessings that her husband had brought them and to focus on her children. Remember, he wrote, "at their age the loss of a mother is much greater than of a father; think what endearing duties you have to discharge." Several months later, when Peter was four years old, Samuel traveled to Geneva to chaperone the family back to England.

The loss of a parent in childhood can have profound effects, including the development of anxiety and depression later in life. In Peter's case, the immediate impact was a peripatetic way of life imposed by his mother. Deeply distressed and unable to feel settled after her husband died, Catherine began moving her children from one town to another, likely hoping that new surroundings would provide a fresh start and heal her pain.

Over the next 10 years, the family of three lived in a multitude of locales, stretching from London's Kensington suburb to Edinburgh, Scotland. The pressure on Catherine to maintain a respectable life and keep her children safe and healthy as a single mother often overwhelmed her, and she was not always able to make the best choices. At one point, Catherine settled the three of them into what she discovered to be "a lowlife neighborhood surrounded with women of bad character."

Biographer Joshua Kendall contends that the constant moves during childhood left Roget feeling insecure and unsure about what was to come; this, in turn, shaped the way he interacted with the people and places around him. "To feel rooted and connected, Peter turned inward—away from the real world," writes Kendall. "He became a daydreamer who easily got lost in the contents of his own mind."

Rather than seek comfort in people, Peter became preoccupied by subjects that interested him, including astronomy and the cosmos. Catherine wrote in a letter to her brother that "Peter, ever eager after new studies, has for this while left this world and lived wholly in the Starry regions"—so much so that he gave his mother and sister a three-hour lecture on the topic, leaving them both thoroughly exhausted.

Within the tumult of his world, language became Peter's safe harbor. In 1787, when he was eight years old, he began filling a notebook with Latin words and their English translations. He then classified the words into subject areas, including "Of the Weather," ("sun," "moon," "cloud"); "In the Garden" ("lettuce," "leaf," "tree"); and "Beasts" ("tiger," "hog," "rat"). He couldn't understand the trajectory of his childhood, but he could order the language that described it.

Just as Peter fixated on words, his mother obsessed about her boy—how he was feeling, what he was interested in, and where he might land professionally. In letters to her brother, she described her son's colds and coughs; his love of mathematics and chemistry; and a mind she believed was ill suited to business. Her daughter, Annette, by contrast, received far less attention. Catherine's letters reveal only that Annette spent much of her time playing piano and drawing while Peter's surrogate fathers—his uncle Samuel and Étienne Dumont, a Swiss intellectual and close family friend from Geneva—mapped out Peter's future.

By the time Peter was in his early teens, the question of what he might become grew more urgent. Catherine routinely agonized over how little money she had and often asked for assistance, which her brother provided. It was imperative that Peter prepare for an established vocation that would provide him with a decent income and a professional future. By the time he was a tall, lanky 14-year-old, the decision had been made: Peter would study medicine at the prestigious University of Edinburgh.

Roget's fastidiousness, a quality that would drive his linguistic work, became increasingly evident as he got older. In a letter he wrote to his uncle during his first term at university, Peter described the basics—the lectures he attended on Roman antiquities, his studies of Latin and Greek (both languages were required for reading medical texts), and the smell of decomposing bodies in anatomy class. But he also punctuated his account with unusual specificity, including the precise times his classes took place, the 320 steps he climbed on a daily basis, and the exact operating hours at the library. "It is open four days in the week, from 10 to 1," he wrote to his uncle, who may or may not have found such detail as interesting as Peter did.

Five years later, in June 1798, Roget graduated with a medical degree. But it would be several years before he sought his own jobs or even practiced as a doctor. Now 19 years old and accustomed to living with his mother and sister, Peter was far from certain about what to do next and relied on his closest advisers to orchestrate his professional path. By then, his uncle Samuel was a successful London lawyer, and Dumont had relocated from Geneva to London, where he worked as a translator.

Roget's early perambulations after medical school landed him at the Pneumatic Institution, where the famed nitrous gas experiments took place, and then to an apprenticeship in London, arranged by his uncle, with an inventor named Jeremy Bentham. Initially excited by the prospect of working with Bentham to build a "frigidarium," an early model of the refrigerator, Roget soon determined that Bentham had no real understanding of the science required—nor the resolve to actually finish the project.

The next few months left Roget feeling unsettled, yet again, by a situation he could not control. Finding a position as a doctor in London required not just a medical degree, but also professional connections in an unfamiliar city where he had few social acquaintances; he was disheartened by the unpredictability of it all. "Not having had a fixed plan, not having seen anything stable or certain for the future," he wrote his mother, "I have not been able to enjoy anything."

These streaks of hopelessness ran through the Romilly bloodline and became a frequent topic of correspondence between Roget and his mother, sister, and uncle. Often, one would despair and the other offer solace or instructions on how to emerge from the darkness. In letters to his nephew, Samuel candidly acknowledged the ancestral affliction. "Despondency is, I have always thought, the great defect of our family, and I do not think you are more exempt from it than the rest," he wrote.

Biographies of Roget reveal the depths of the family's emotional strain. A first comprehensive volume, published in 1970 by Donald Emblen, offers a rich account of his life and the meanderings of his mind. Kendall's *The Man Who Made Lists,* published in 2008, magnifies the psychological lens, focusing on the long-term significance of Roget's childhood and the looming legacy of family members who struggled with mental illness.

Peter's maternal grandmother, Margaret, was likely overwhelmed by a mood disorder, coupled with grief after seeing six of her babies die. She was "incapable, from the bad state of her health," in the words of her son Samuel, to look after her surviving children. Both Peter's mother, Catherine, and his uncle Samuel, who were raised by their father along with a family relative and a loving servant, suffered the impact of their mother's depressive condition. Samuel acknowledged his own depression, which would later end his life. Catherine, in turn, was anxious, moody, and prone to paranoia; Annette suffered so many ills, both physical and psychological, that her mother tried a slew of remedies, from cream of tartar and rhubarb to port wine.

In Kendall's estimation, Roget's preoccupation with words and the attention he paid to details and numbers—including those 320 steps he counted at university—suggest he likely had what we now know as obsessive-compulsive disorder or OCD. This would also account for his lifelong attention to dirt and hygiene. The streets in Paris, he would later observe, were "greasy" and covered in mud. And he could barely stand the filth and air pollution in the town of Manchester, where he later began his medical career.

Ancient philosophers had theorized that genius could not exist without a touch of madness. Although there's no scientific consensus on a link between mental illness and great achievement, history offers abundant examples. Sergey Rachmaninoff stopped composing music for almost three years during a severe bout of depression. Virginia Woolf wondered in her diary about how to describe "the violent moods of my soul," even as she wrote more than a dozen books and hundreds of essays and reviews. Vincent van Gogh, who had psychotic episodes, painted "The Starry Night" from inside a mental health institution before taking his own life at the age of 37.

Mental health conditions exist on a spectrum and manifest in different ways. In some cases, a quality that debilitates one person can enrich the creative life of another. Robert Lowell, who had bipolar disorder, called his shifting moods "pathological enthusiasm" and "dust in the blood." The energy his mania triggered allowed him to write frenetically and productively, his biographer Kay Redfield Jamison contends, before depression sapped him to his core.

Greta Thunberg, the Swedish teenager who began leading an international campaign to raise awareness and tackle climate change when she was 15 years old, speaks openly about her Asperger's syndrome (now classified as autism spectrum disorder) and claims it as a benefit. In a crisis as big as global warming, she said in an interview on *CBS This Morning*, "we need people who think outside the box and who aren't like everyone else."

Asperger's has emerged in a similar vein in analyses of some of the world's most brilliant scientists. Reports have suggested that behaviors common to both Albert Einstein and Isaac Newton—among them, a solitary demeanor and intense focus on a particular subject—line up with features of autism. Henry Cavendish, the 18th-century chemist and physicist who discovered hydrogen, was withdrawn, extremely literal, single-minded, unconventional, and obsessed with number crunching. The neurologist Oliver Sacks wrote that these qualities are evidence of Asperger's and also "the very traits he used so brilliantly in his pioneering scientific research."

And Bill Gates, whose rocking behavior has inspired public queries about whether he might be on the autism spectrum, has touched on his own identification with at least one of the condition's features. In a 2014 book review of the novel *The Rosie Project* by onetime information technology consultant Graeme Simsion, Gates wrote: "Anyone who occasionally gets overly logical will identify with the hero, a genetics professor with Asperger's Syndrome who goes looking for a wife. (Melinda thought I would appreciate the parts where he's a little too obsessed with optimizing his schedule. She was right.)" Gates also made a point of embracing that trait. "It's an extraordinarily clever, funny, and moving book about being comfortable with who you are and what you're good at."

Whether or not Roget had OCD, he clearly thought differently, and his fixation on words led to his magnum opus. But his uncle warned him not to let emotional upheaval destroy him. "I don't tell you that I could wish you to be a little more cheerful than you are because I know that gaiety is not to be put on like one's coat," he wrote, "but I know by experience that melancholy may be very much increased by being indulged."

⌒

What the young medical school graduate needed to buoy his mood was a change of scenery and a road map. Instead of nailing down a medical job, as might have been expected of a young M.D., he left the country as chaperone and tutor for the teenage sons of John Philips, a wealthy cotton mill owner, and his wife, Margaret. "At last he was about to embark on a project of considerable responsibility and very lively interest," his biographer Donald Emblen wrote, "and to be thrown at last on his own resources, with no anxious mother or uncle hovering at his shoulder."

Over the next year and a half, from February 1802 to November 1803, Roget and the Philips boys traveled from Paris through east-central France to Geneva, the last place he had seen his father alive. Along the way, unsurprisingly, they documented numbers. In a letter home, Burton Philips noted 360 steps to Notre Dame's tower and 3,800 organ pipes, and the four-mile length of the Bois de Boulogne (teeming with visitors between 3 and 6 p.m.). His brother Nathaniel, meanwhile, reported that there were 980 paintings and 209 statues and busts displayed at the Louvre.

Now in his early 20s, Roget was forced to mature—answerable not only to himself, but also to his two young charges and their parents. Every day he was tested as they traveled, but never more so than when the trio found themselves in Geneva in May 1803, just as Britain declared war on France. At the time, the city was under Napoleon's rule and all Englishmen over the age of 18 were ordered to be taken prisoners of war. Relying on quick thinking and old family connections, Roget shrewdly managed to get himself and the boys safely back to England.

Spark

Once home, Roget finally landed his first full-time job in the field in which he had trained. At 25—about the same age his countryman and forebear Isaac Newton was when elected a Fellow of Trinity College at Cambridge—Roget became a physician in Manchester, an industrial city in northwest England plagued by public health challenges.

Although their professional trajectories moved at different paces and in different directions, Newton and Roget had something in common: a childhood passion that never left them. In Manchester, after wrapping up his daily doctoring hours, Roget turned back to language. By 1805, the 26-year-old had created his 15,000-word compendium. Although imperfect, it would become a valuable first draft (an eager bidder would buy it at auction in 1992) of the version he would complete five decades later.

In his late 20s and 30s, Roget began taking charge of his personal and professional life, which he expanded to encompass interests beyond clinical medicine. In addition to practicing as a doctor, he wrote articles for the *Encyclopedia Britannica* and began lecturing at the Royal Institution in London, where he had since relocated. He engaged in intellectual dialogues with the likes of celebrated scientist Michael Faraday and, in his off hours, invented a logarithmic slide rule—used to calculate the roots and powers of numbers—which earned him the coveted title of Fellow in London's Royal Society at the age of 36.

Nevertheless, trauma continued to put a strain on Roget, marking midlife with two unimaginable losses. In October 1818, when he was 39 years old, his uncle Samuel's wife, Anne, died after an illness. By then an accomplished legal reformer and esteemed member of Parliament, Samuel broke down, entering into what the *Times* of London reported as "a state of anguish which approached to frenzy." Summoned a few days later, Roget found Samuel bleeding heavily over a washbasin in his bedroom. He had slit his throat with a razor and, according to biographer Emblen, "collapsed in the arms of Roget, who lowered him to the floor."

In the years following his uncle's suicide, a distraught Roget continued to practice medicine, taking on various clinical responsibilities, including physician to the Spanish Embassy in London. But he fretted over the pace of his career, despite his achievements. "I think I hear you asking me how I go on professionally—I should answer *tolerably*," he wrote in a letter to his closest friend, a fellow scientist named Alexander Marcet in March 1821. "I do not feel that I am advancing; nor perhaps am I receding."

Not even two years later, Roget experienced another painful loss when Marcet died suddenly after collapsing in his home. As with his uncle, Roget had been summoned to the scene—Marcet's wife called him in desperation—but could do nothing to save his friend.

The tragedies in Roget's personal life seemed unforgiving, but there were also periods of joy, most notably when he fell in love with a young woman named Mary Hobson, the only daughter of a well-off merchant family from the town of Liverpool. It is unclear how the two met— possibly through a mutual Liverpool friend—but their relationship evolved at a time when Roget was eager for companionship. By then a middle-aged bachelor who had endured several failed romances and witnessed the fragility of life, he was ready to settle down.

In November 1824, the 45-year-old Roget married Mary, 29. The couple had two children, Catherine and John, and began a contented life filled with visits to the theater and long walks in the family's Bloomsbury neighborhood. But less than a decade into their marriage, Mary was diagnosed with cancer and died at the age of 38.

Roget—now a 54-year-old widower well into the latter half of his life—could do only one thing: Find sustenance in his work. A highly regarded lecturer, he was appointed the first Fullerian Professor of Physiology at the Royal Institution (the organization credited Roget with "extraordinary tact and ability"), continued serving as secretary to the Royal Society (a position to which he had been appointed in

1827), and completed his greatest scientific undertaking: a surveillance of the vegetable and animal kingdoms, published in 1834. In two volumes of more than 1,000 pages, the churchgoing Roget balanced views of creation (it would be another 25 years before Darwin's *On the Origin of Species* would be published) with biological explorations of everything from the spongioles on the tips of plant roots to nerve fibers found in the abdomens of lobsters.

Roget's treatise firmly established him as not only a preeminent scientist, but also as a meticulous organizer whose skills could be transferred to entirely other subject matter. Over the next few years, he remained active in scientific affairs until he closed his clinical practice in 1840 and resigned from the Royal Society eight years later. "Having now grown grey in that service, I feel that it is time for me to retire, while my strength is yet unbroken," he wrote.

It was time to return to the childhood passion he had tucked away for almost 50 years.

How many of us pursue our childhood interests? We all have them— bug collecting, cartoon drawing, clarinet playing—though sometimes we forget they existed. This occurred to me one morning as I sat in the auditorium at National Geographic Society headquarters with hundreds of middle school students who had come to learn from Ryan Carney, assistant professor of digital science at the University of South Florida. Carney made a point that stuck with me: We should pay attention to our childhood dreams.

A National Geographic Emerging Explorer, Carney works at the intersection of biology and technology. On a giant screen, he showed an animated hologram of *Archaeopteryx,* the first dinosaur known to be capable of flight. Carney created his visual by taking X-ray scans of fossil bones, which he then manipulated with computer animation and virtual and augmented reality. Carney's dynamic model allows paleontologists to study the *Archaeopteryx* in 3D, as if the dinosaur had come back to life.

I was mesmerized by Carney's work, but even more bowled over by a series of images he showed: his own childhood drawings of dinosaurs in flight. When I called Carney later to learn more about his professional journey, he told me that he had always been interested in paleontology, but it wasn't until recently that he realized how closely aligned his early hobby was to his current career.

As he was preparing for a lecture a few years ago, Carney's mother sent him a series of sketches he'd made when he was six years old, which she had dug up at home. Now in his mid-30s, Carney had not seen the drawings in three decades and he was startled by the image in one of them: a dinosaur with a single isolated feather. The drawing was almost identical to a sketch he made of an *Archaeopteryx* for the title page of his dissertation some 25 years later. "It sent shivers down my spine," he told me.

Today, Carney makes a point of showing students the drawings he made when he was six years old because he wants to be sure they never lose sight of their passions. "I think what inspires you as a child may inspire you as an adult," he says. "That was one of the keys to my success."

Too often, the dinosaur dreams and astronaut fantasies of childhood are deemed joyful diversions, not lifelong aspirations. Surveys show that most people do not pursue their childhood ambitions—they get tossed aside as we age. Carney managed to seize his and move forward. Roget was another breed: a rediscoverer who resurrected a young love in the final phase of his life.

In 1849, when Roget was 70 years old and settling into retirement, he picked up his 1805 draft of word lists and got to work. Others had published books of synonyms before Roget, but none could compare in breadth and specificity. Although Roget modestly described his old manuscript as "scanty and imperfect," the compilation had served him well as a literary companion over the years, and he believed it would prove useful to others as well.

Roget was not the likeliest word enthusiast, as his biographer Emblen makes clear. Trained as a scientist, he did not read much beyond the scholarly texts that lined his library. And his writing was far from elegant, as even a brief note to a newlywed friend makes clear: "Our pleasure has been greatly heightened by the very favourable accounts we have heard of the character and accomplishments of the amiable object of your choice." Critics described Roget's writing as "ponderous," "stiffly formal," "priggish," and "sententious."

Still, his compulsion to organize superseded all else to become his greatest asset. Drawing on the work of renowned Swedish botanist Carl Linnaeus, the 18th-century taxonomist who organized nature into domains, kingdoms, phyla, classes, orders, families, genera, and species, Roget mapped out six groupings for his collection of words: Abstract Relations, Space, Matter, Intellect, Volition, and Affections. Under each of these, he listed like words in numbered categories. He called his compendium a thesaurus after the Latin word for "treasure" or "treasury."

The first edition of Roget's *Thesaurus of English Words and Phrases* was published in 1852 when Roget was 73 years old. Although some were initially puzzled as to how to use it ("a curious book," one reviewer noted), it was clear that the depth of Roget's work was extraordinary and his offerings vast. The book was neither a dictionary nor simply a compendium of synonyms; instead, the inveterate organizer had sorted and classified *"all human knowledge,"* as Emblen enthusiastically emphasized.

Within his six overarching categories, Roget presented a total of 1,000 subheadings. (The first edition had 1,002, but Roget later reduced it to a tidy 1,000.) Under "Intellect," for example, Roget provided a range of words that included *imagination* and *dupe*—seemingly dissimilar but connected by concept. An anonymous reviewer writing for the *Westminster Review* in 1853 compared Roget to the British lexicographer whose groundbreaking *Dictionary of the English Language* debuted a century earlier: "Roget will rank with Samuel Johnson as a literary instrument-maker of the first class."

The initial 1,000 copies sold out speedily, and the book soon collected well-known devotees, including British author and playwright J. M. Barrie, who tipped his hat to Roget in *Peter Pan:* "The man is not wholly evil," he wrote in describing Captain Hook, "he has a Thesaurus in his cabin." Barrie later signed the title page of his own 1867 edition with a note: "with the help of this copy of the Thesaurus I wrote all my books and plays."

In the United States, the emergence of the crossword puzzle, first published in *New York World* in 1913, gave the thesaurus a newfound raison d'être. Sales spiked as fans found a literary companion they could reach for when a six-letter word beginning with D eluded them. "Here were more golden words than even the emptiest mental purse required," the *New York Times* declared in a full-page article on the subject in February 1925. The story's headline spelled it out perfectly: "Roget becomes Saint of Crosswordia; His 'Thesaurus' Not a Monster of the Reptile Age, but a Treasury of Synonyms."

American writers Sylvia Plath and Maya Angelou cherished their copies like literary paramours. Plath said she'd bring her well-worn volume to a tropical island and told her boyfriend Gordon Lameyer, "I love you more than the alphabet and Roget's thesaurus combined." In a 2013 interview, Angelou cited the compendium as one of her must-have writing tools: "I have a *Roget's Thesaurus,* a dictionary, a Bible, a yellow pad, and pens, and I go to work."

Roget chose to group his words in categories, not alphabetically, because he intended it for people who understood language but needed a hand finding the appropriate word. It was decidedly *not* a dictionary, but a catalog that would prod the memory and illuminate a subject, "like a beam of sunshine in a landscape." Just as actors need wardrobes and painters need colors, a writer requires a "copious vocabulary," he wrote in his introduction.

The list is so rich and informative that plenty of word lovers have read it for pure enjoyment. George Douglas, an emeritus English professor at the University of Illinois, described how looking up "punishment," for example, led him down a delightful pathway of words and phrases that included fustigation, keelhaul, and gibbet. "A juicy set of words," Douglas wrote. "I doubt if I am ever going to use *argumentum baculinum*, or 'rub down with an oaken towel,' but somehow I take delight in letting Roget feed me more words than I really need."

Despite Roget's introduction and subsequent editors' instructions on how to use the book, it has long suffered from misuse by novice writers. In his memoir, President George W. Bush described pulling out the copy his mother had tucked into his luggage as a newly arrived Texan boy at Phillips Academy Andover. In one of his first writing assignments, Bush chronicled the sadness he felt after his younger sister, Robin, died of leukemia at the age of three, writing: "Lacerates were flowing down my cheeks." His teacher was appalled, and Bush was humiliated, he recalled in his memoir. "When the paper came back, it had a huge zero on the front."

Countless college students and their educators can attest to similar fumbles, but it wasn't until journalist Simon Winchester penned a lengthy takedown of *Roget's Thesaurus* in the *Atlantic* in 2001 that the book took on its own form of celebrity scandal. In his piece, Winchester depicted his own unfortunate run-ins with *Roget's* as a professor—a student changed "earthly fingers" to "chthonic digits" after consulting the thesaurus—and argued that it "has become no more than a calculator for the lexically lazy."

"Word Imperfect," as the article was titled, provoked a range of responses, including a letter to the editor from a Roget defender in Cibolo, Texas, who turned Winchester into a thesaurus entry of his own: "Winchester: Arrogant, presumptuous, elitist, didactic, verbose, iconoclastic, impractical."

Since the first edition debuted, *Roget's Thesaurus* has been updated multiple times in British and American editions that have sold some 40 million copies worldwide. Lexicographer Barbara Ann Kipfer has compiled dozens of books about words and edited *Roget's* U.S. edition for the last three decades. She spent more than a year preparing the most recent edition for publication. Among her many tasks: checking word lists for accuracy, removing obscure language, updating spellings, and reminding readers, yet again, that a thesaurus does not replace a dictionary, but rather complements it. In her introduction to the new edition, Kipfer underscores the significance of Roget's work: "The revolutionary achievement of Dr. Roget was his development of a brand new principle: the grouping of words according to ideas."

Kipfer's enthusiasm for words is palpable when we talk. She loves dipping into language and takes delight in figuring out how to make *Roget's* as comprehensive and relevant as it can be in the 21st century. Among the roughly 443,000 words and phrases in the newest edition are 2,400 that appear for the first time. These include "abso-bloody-lutely," "Asian flu," "bada bing bada boom," "bromance," "faceplant," "internet café," "neuroplasticity," "reality show," and "yuppify."

I couldn't help but ask Kipfer about a word that is especially pertinent to Roget: "late bloomer." The word first appeared in the *Oxford English Dictionary* in 1827, she tells me, but only in reference to plants that flowered in late summer or early fall. It was not until 1887 that a figurative definition was added: "a person who displays talent, develops skills or interests, or achieves success at a relatively late stage." Roget did not include the word in his early editions, but it does appear in the fourth edition, published in the United States in 1977. Under the general category, "Lateness," "late bloomer" is listed as a synonym for "latecomer," along with "Johnny-come-lately," "dawdler," "dilly-dallier," and "slug-abed" (which means exactly what it sounds like).

It seems fair to say that late bloomer does not tend to have the rosiest connotation—at least not in relation to the synonyms it's clustered with—nor does it boast a positive social or professional reputation. In the increasingly fast-paced world we inhabit, it seems almost passé—in the "out" column, along with pipes and corsets. Today, we celebrate wunderkinds: young scientists, billionaires, athletes, CEOs, chefs, and entertainers, as if the pairing of youth and achievement is the brass ring we all should seek.

Nevertheless, late bloomers have a rich history. In 1930, novelist Harriet Doerr left Stanford University before graduating to get married. At the age of 65, almost 50 years later, she went back to complete her degree. At 69, she received Stanford's prestigious Stegner Fellowship in fiction and studied with writer John L'Heureux (who started writing poetry during his first career as a Jesuit priest). He helped her organize several of her short stories into her breakthrough novel, *Stones for Ibarra*, published in 1983. The book, which a *New York Times* review credited with "echoes of Gabriel Garcia Marquez, Katherine Anne Porter and even Graham Greene," won a National Book Award for First Work of Fiction.

Doerr's semiautobiographical account of a couple living in Mexico, published when she was 73 years old, was based on her travels to the country with her husband to oversee his family's mine business. Memories saturate her prose: "the width of the sky, the depth of the stars, the air like new wine, the harsh moons and long, slow dusks." Doerr could not have written the novel at a younger age; she had to live it first.

"Although Harriet Doerr had come to writing very late in life," Wallace Stegner wrote in an appreciation published in *Esquire* in 1988, "she discovered, as we all did, that she was an almost flawless lens, with a capacity to make a world out of the fragmentary images she had caught."

For many women, including Doerr and the artist Grandma Moses, inspiration surfaces after the children are grown. For some, the muse alights in its most noble form late in life, when the mind has time to flourish. Hungarian composer György Kurtág completed his first opera, *Samuel Beckett: Fin de Partie,* at the age of 91, earning rave reviews from critics after its 2018 debut in Milan. "There are many musical monuments of old age: Monteverdi's *The Coronation of Poppea,* Verdi's *Otello* and *Falstaff,* Janáček's *From the House of the Dead,*" wrote the *New Yorker's* Alex Ross. "But no composer has ever taken so huge a leap so late."

In his 2019 book, *Late Bloomers,* Rich Karlgaard, publisher of *Forbes* magazine, laments society's obsession with young superachievers. Encouraging and celebrating early success is not the problem, Karlgaard writes. (As it happens, his own magazine publishes an annual "30 Under 30" list, extolling the accomplishments of wunderkinds.) But he calls for an end to glorifying precocious accomplishments. This fixation on youth undercuts the potential of late-blooming discovery and undervalues human qualities that grow with time, he writes. "It trivializes the value of character, experience, empathy, wisdom, reliability, tenacity, and a host of other admirable qualities that make us successful and fulfilled."

Late bloomers, whether first-timers like Harriet Doerr or rediscoverers like Roget, have another advantage in age: They have lived in different realms, allowing them to pick from different baskets of experience rather than be narrowly limited to one. Doerr's life as a wife, mother, traveler to Mexico, and 60-something college student, permeated the lens through which she wrote. Roget's experience as a doctor, scientist, and middle-aged husband and father informed the way he observed and organized the world.

Roget perfected this ability in every realm. A simple glance through the window could become a science experiment, as it did one morning in 1824, when he looked through the slatted blinds in his kitchen. Roget noticed that the spokes of a carriage wheel passing by outside appeared to be curved—an optical illusion he attributed to the eye's ability to see a sequence of still images flow uninterrupted. A report he wrote for the Royal Society later contributed to a field Roget would never witness: moviemaking.

In the preface to the first edition of his thesaurus, Roget pointed out that despite his incessant dedication to the project, his compilation was far from perfect. "Notwithstanding all the pains I have bestowed on its execution," he wrote, "I am fully aware of its numerous deficiencies and imperfections, and of its falling far short of the degree of excellence that might be attained."

It was, however, finished, and there would be plenty of time for perfection as future editions rolled out. In the words of Russian novelist Ivan Turgenev: "If we've to wait until everything, absolutely everything, is ready, we shall never make a beginning."

⌒

The Bloomsbury section of London has long been home to residents who write. Edgar Allan Poe spent a few influential years here as a child in the early 1800s. Charles Dickens wrote *Oliver Twist* and *Nicholas Nickleby* at 48 Doughty Street in the late 1830s. J. M. Barrie, author of *Peter Pan,* lived on the corner of Bernard Street. Virginia Woolf; her husband, Leonard Woolf; and E. M. Forster were among the writers who gathered in their Bloomsbury homes to discuss their creative pursuits in the early 20th century.

As I walked through the area one hot August day, it became apparent that Roget would have recognized little about the place where he spent so much of his life. The bustling neighborhood is now a galli-maufry of old and new—terraced houses, Royal Mail post boxes, taquerias, pedestrians glued to their cell phones, and buses filled with tourists speaking Chinese.

Still, there was something sublime about ambling down Roget's old street, Upper Bedford Place (now called Bedford Way), just off leafy Russell Square. As I did, I couldn't help but think of my visits to the lively square outside of Picasso's childhood home in Málaga, Newton's bedroom in Woolsthorpe Manor, and Eleanor Roosevelt's beloved Val-Kill cottage—all soulful remnants of great minds that endure.

Roget lived the last 20 years of his life in his Bloomsbury home with his daughter, Kate, who had never married. Here, he compiled the final draft of his thesaurus, which sold for 14 shillings. As word of his achievement grew, demand for the book increased. Roget over-saw a staggering 28 additional editions in his lifetime—so many that the repeated printings wore out the stereotype plates that had been created for the third edition in 1855.

In the summer of 1869, when he was 90 years old, Roget traveled with Kate to the town of West Malvern in the hills of Worcestershire. Roget and his daughter had retreated to the village for years, and there he died on September 12, while on holiday. After a service at St. James Church, Roget was buried in the peaceful churchyard, far from the busyness of London.

After his death, Roget's son, John, a lawyer who found his own true passion in watercolor painting and history, discovered a copy of his father's thesaurus with words and phrases scribbled in the margins that he had intended to add. John went to work, tweaking and editing, and published an updated edition in 1879. John's son, Samuel Romilly Roget, took over as editor in 1925.

Before he launched into the last two prolific decades of his life, Roget had mused about the process of aging. "Every living being has a period assigned for its existence," he wrote in the *Cyclopaedia of Practical Medicine* in 1845. "With the germs of life are intermixed the seeds of death; and, however vigorous the growth of the fabric, however energetic the endowments of its maturity, we know that its days are numbered."

All true. But as Roget, the late bloomer, proved better than anyone: The energetic endowments of maturity can give rise to an everlasting legacy.

CHAPTER 12

Grandma Moses

The Painter (1860–1961)

Hoosick Falls, a modest village with a mighty claim to fame, sits in New York's emerald green hills near the border of Vermont. Years ago, the area was home to factories making farm machinery, paper, and glass along the rolling Hoosic River. Today, American flags hang from iron lampposts lining Main Street, potted flowers brighten sidewalks, and 19th-century architecture mingles with chai lattes and kombucha tea.

Here, in April 1938, a historic moment transpired—an occurrence that happened by chance when a traveling hobbyist art collector from New York City, Louis Caldor, noticed a display of paintings in the window at Thomas's drugstore on John Street. The artworks, depicting nostalgic and colorfully whimsical vistas of farm life, had lingered in the shop for months and attracted little attention. But something about them—the subject matter? composition? colors?—caught Caldor's attention, and he asked if there were any others to see. By the time he had walked out of Thomas's, Caldor had bought up every one of them.

The painter was Anna Mary Robertson Moses, a resident of the nearby town of Eagle Bridge. Then a 77-year-old widow, Moses had taken up art, a childhood amusement, to keep herself busy after her farming days had ended and her children had grown. She had no serious expectations that her paintings would sell. Not long before Caldor's visit to Hoosick Falls (the town took its name from the river, but added the "k"), Moses had taken a sampling of her art to the Cambridge Fair seven miles up the road. Her canned fruit and raspberry jam won prizes; her artwork did not.

But Caldor was intrigued—so much so that he made a point of stopping by Moses' house, set amid fertile farmlands and sweeping skies, where her daughter-in-law Dorothy greeted him. In a memoir Moses wrote years later, she recalled the message Dorothy gave her when she came home that day: "If you had been here, you could have sold all your paintings. There was a man here looking for them, and he will be back in the morning to see them."

In preparation for his visit, Moses hastily gathered her paintings. Dorothy had told Caldor that Moses would have 10 paintings to show him—but Moses soon discovered that she only had nine. Eager to increase the lot, she took one of her larger works, a painting of the Shenandoah Valley, and cut it in two. Then she patched up the sides and added a second signature.

Grandma Moses, as the artist came to be known, was a pragmatic, feisty, and hearty soul whose destiny was molded not by choice but by circumstance. Born on a farm near the village of Greenwich, New York, and subsequently married to a farmer herself, Moses' days were filled with chores, motherhood, and the bounties of nature. Her life straddled two centuries, bookended by the Civil War in the 19th and the Vietnam War in the 20th.

It was not until her 70s that Moses picked up a brush and started to paint. Her studies of rural landscapes depicted a way of life that charmed those seeking simplicity—a time when horses drew carriages through the snow, farmers chopped wood, and children delighted in playing on seesaws. One month after her 80th birthday, the artist's first solo show debuted at a Midtown gallery and her work soon attracted the attention of art critics and a generous following of fans. Some of them trekked to her home in rural New York to get a peek of their beloved painter with her impish eyes and smile.

Between the late 1930s and her death in 1961, Moses sat at her flat pine wood table turning out tableaus of rustic serenity as the world churned with its second great war and the instability that followed. At the age of 88, she had tea at Blair House with Mrs. Truman and the president, whom she cajoled into playing a command performance on the piano. At 95, she gave an interview to Edward R. Murrow. By 100, she had completed more than 1,000 paintings and seen her work reproduced on fabrics, wallpaper, and Christmas cards, which sold by the millions across the country.

Anna Mary Robertson Moses is the quintessential late bloomer, a woman whose singular career kicked off not only belatedly but in the eighth decade of life. By then, the artist had outlived her life expectancy at birth by more than 30 years. Moses' remarkable longevity allowed her to hone a hobby into a professional career. Unlike prodigies and midlifers, whose peers stand witness to their achievements, Moses' talent emerged at an age that most of her contemporaries never lived to see.

As she celebrated her 100th birthday, Moses pondered her life with typical unsentimental clarity. Antiquity is what you make of it, she told her friends and family. "With an interest in some form of work and with a reasonable appreciation of what is going on in the world, old age can be a joy," she said. "I'm 100, but I feel more like a bride.

Now that I'm here, the only thing to do is to go back to the beginnin' and start all over again—and that's what I'm gonna do."

⌣

"I, Anna Mary Robertson, was born back in the green meadows and wild woods on a farm in Washington County, in the year of 1860, September 7th, of Scotch Irish ancestry." Thus begins Moses' autobiography, *Grandma Moses: My Life's History*, published the year she turned 92. Moses opens and ends her remarkable account in the rural environs of upstate New York, a place cocooned in snow in the winter and dappled with whip-poor-will blossoms in the spring.

In her memoir, Moses documents the arc of her life in the same way she painted, splashing literary panoramas with flecks of vivid color. Her ancestors, we learn, settled early in Washington County, where one of them married a Native American. Her great-grandfather on her father's side joined Albany County's 16th regiment to fight the British at Fort Ticonderoga in 1777. As a child, her maternal grandmother used to sneak bread out of her aunt's shop to give to a poor family across the street—until the day a talkative parrot tattled on her.

It is unclear how Anna's father, Russell King Robertson, met her mother, Margaret Shanahan, or when they were married. But over 20-odd years, the couple produced 10 children—five boys and five girls. Anna, the third, was sandwiched between two older brothers and one younger, all born in quick succession. A second quartet arrived next, followed by a lull before the final two. "We came in bunches, like radishes," Moses wrote.

Moses delighted in the stories of her family roots and was especially proud of her own frugality, a trait she attributed to her Scottish heritage. In her memoir, she succinctly documented the characteristics she inherited through both her paternal (the Kings and Robertsons) and maternal lines (Shanahans and Devereaux). "From the Kings I got art, from the Robertsons inventive faculty, from the Shanahans thrift, the Devereaux generosity," she wrote. She was also an adventurous, commonsensical tomboy who couldn't sit still.

The first 12 years of Moses' life read as a blissful mix of family farm duties and sibling merriment. There was always something to do—a baby sister who needed rocking, sewing lessons from her mother while her father worked at the flax mill, the sporadic schooling allowed for girls. Much of Anna's time was spent outdoors, even when the temperatures plunged below zero. She loved collecting sap from maple trees, heating it into syrup, and smothering it on buckwheat cakes and hot biscuits. One of her most beloved pastimes was riding in a horse-drawn red sleigh as sprays of snow fell on top of her from the hemlock trees. "Oh, those happy days!" she wrote.

Although Moses did not dedicate herself to painting until much later in life, making art was a form of family entertainment early on. Her father, whom she described as "a farmer and inventor, a dreamer, a believer in beauty and refinement," painted murals on the walls of the house and bought his children sheets of paper for a penny apiece so they could draw. Anna brightened her pictures with pigment from grape juice and berries and used indigo to color the eyes of her paper dolls. At one point, she began painting "lamb-scapes"—as her brothers joked that she called them—on sticks of wood and pieces of slate.

The artwork she made as a child wasn't bad, she recalled her father saying, "but mother was more practical, thought that I could spend my time other ways." Unlike her younger contemporary Picasso, whose artistic life seemed fated from toddlerhood, Anna's destiny was decreed largely by her gender. Social norms dictated that women stay home to care for children—and although glimmers of change were beginning to emerge, women's suffrage would not be ratified until almost half a century later in 1920, when Moses was 60 years old.

Spark

When she was 12, Anna left home to earn money as a hired hand for families and elderly neighbors. In her memoir, Moses describes this period of her life as a mix of satisfaction and hardship. On top of the required work—cooking, ironing, washing, sewing, and hoeing gardens—she had the opportunity to mature and mingle with the outside world. But she was also young and scared; she later said she thought one of the houses she lived in was haunted. And she suffered tragic personal losses in her family. When she was in her late teens and early 20s, an older brother and a younger brother and sister died from respiratory illnesses, including the adverse aftereffects of measles.

Moses' parents imbued her with a pragmatic acceptance of life's tribulations—"as you are born you must die," her mother said —and it was an outlook she sustained throughout her life. Rather than dwell on misfortune, she accepted it and embraced simple joys where she could find them, taking pride in serving guests cured beef with hot biscuits and delighting in the fresh smell of flowers at the state fair. "We had to take the bitter with the sweet always," she wrote.

In the fall of 1886, the year she turned 26, Anna met Thomas Salmon Moses, a hardworking man with a good reputation. Anna never seemed bothered by how old she was at the time, which would have been considered advanced by conventional standards of marriage; instead, she focused instead on the couple's commonsensical compatibility. "He found me a good cook, and I found him of a good family, very temperate, and thrifty," Anna recalled. After a yearlong courtship, the couple was married late one November afternoon in Hoosick Falls, before heading south to warmer weather. The groom wore black; the bride, a dark green dress with multiple corsets, high-buttoned shoes, and a hat with a pink feather. Just after completing their vows, they rushed to make a six o'clock train out of town, skipping their supper and wedding cake.

At 27, Anna entered a new phase of life as a young adult, farmer's wife, mother, and income earner in the hills of Virginia's Shenandoah Valley. Over 15 years, she gave birth to 10 children; five survived, four were stillborn, and one died at six weeks. She tended to farmwork—sewing, cleaning, cooking, and caring for chickens, cows, and horses. And, to earn extra money, she made butter pats, stamped with "Moses," which she sold for 50 cents a pound. "I could have been a successful dairy farmer had I stuck to it," she later said.

In the winter of 1905, when she was 45 years old, Anna and Thomas loaded up a railroad car with their belongings and relocated back to New York, to the town of Eagle Bridge. Here, in a white farmhouse steps from the Hoosic River, Moses' transformation from farmer's wife to artist would eventually take place. She dabbled at first, making small paintings as Christmas gifts, and completed her first large image—a painting of a sunlit lake and two butternut trees—to cover a board in the parlor fireplace when she ran out of wallpaper.

From the start, purpose, rather than passion, fueled the artist. After her husband died in 1927, when she was 66, her son Hugh and his wife, Dorothy, took over the farm. Moses needed to fill her time. She took up art as a pastime, at first stitching pictures out of brightly colored yarn for a grandchild. But painting was more appealing for two reasons: Moths didn't devour the artwork, and brushes were easier than needles on her stiff arthritic hands. In her 70s, she sat down at her table and became a painter.

Moses had a daily routine, rising early in the morning, taking breaks for a cup of coffee and lunch, and working into the afternoon. Propped up with two pillows on a swivel chair, she laid a fresh piece of pressboard on her table; she preferred it to canvas, she said, because it was stronger and lasted longer. She covered the board with linseed oil and three coats of flat white paint, picked up her paintbrush—the smaller, the better—and started to paint, often working on four images at a time. "My people were Scotch, you see, and by working on four at one time I save paint," she told the journalist and artist S. J. Woolf, who profiled her in the *New York Times* in 1945.

Moses' artistic career could never have happened without the freedom from responsibilities provided late in life. Over the first seven decades, she had been too absorbed by work and family to do anything else. "I never gave any thought to pictures, for I did not have the chance," she said. "I started in and found that it kept me busy and out of mischief."

Moses' odyssey is a story shared by women throughout history—and within these pages—whose lives were preordained by society's expectations and economic realities. Professional and creative endeavors, if they happened at all, almost always transpired after marriage, child rearing, and domestic responsibilities.

The botanist Mary Gibson Henry, born in Jenkintown, Pennsylvania, fell in love with wildflowers on a trip to Moosehead Lake, Maine, in 1884, when she was seven. But it wasn't until her physician husband encouraged her to pursue her passion—after caring for their five children—that she set out on her first plant expeditions in her late 40s.

Over the next three and a half decades, Henry completed more than 200 trips, from the Rockies to British Columbia and Bolivia, to research and collect botanical specimens. She died at 82 while on a collecting trip in North Carolina. By the end of her life, she had written more than 100 journal articles; served in leadership roles, including as director of the American Horticultural Society; and donated thousands of plant specimens to botanical societies, nurseries, and arboretums. "For the first 19 years of my marriage," she said, "my only gardening was in the clouds."

Laura Ingalls Wilder might never have written her beloved *Little House* series if her daughter, Rose, hadn't jump-started her literary career when Wilder was in her early 60s. Acting as Wilder's editor, Rose persuaded her mother to document her prairie childhood— sometimes in an overbearing manner, biographers later discovered— and connected her to the publishing world. In 1932, *Little House in the Big Woods* debuted when Wilder was 65 years old. She hesitated to write a sequel, but could not resist the children and mothers who flooded her with letters wanting to know what happened next. At the age of 76, Wilder completed the last of her eight books, *These Happy Golden Years.*

After Louis Caldor made his consequential stop at Hoosick Falls, he headed to New York City with his acquisitions. Despite his enthusiasm, Caldor initially had trouble ginning up interest in Moses' work. Even if a painting seemed promising, dealers found themselves apprehensive about investing in a self-taught artist of a certain age. "Now and then he was told that the works he showed were quite 'nice,' but unimportant and not in line with contemporary art," the art dealer Otto Kallir later recalled, "above all, nobody wanted to waste time and money in promoting an unknown artist who was close to 80."

But Kallir, the owner of Galerie St. Etienne on West 57th Street, was the exception. Recently arrived from Europe, he was seeking contemporary art with an authentic American sensibility and was therefore willing to take a serious look at Moses' paintings when Louis Caldor showed up in the fall of 1939. Kallir was especially impressed by a piece titled "Bringing in the Maple Sugar." "While the figures were done in a rather clumsy fashion, the landscape was painted with astonishing mastery," Kallir later noted in a biography he wrote of the artist. Eager to see more of her paintings, Kallir met Caldor at his car, which was parked in a lot about a half hour away and served as a storage unit. A pile of Moses' artwork sat in the trunk.

Spark

On October 9, 1940, as the Battle of Britain raged and the Second World War entered year two, Otto Kallir presented "Works by Mrs. Anna Mary Moses," in her first solo exhibition, "What a Farm Wife Painted," at Galerie St. Etienne. At a special preview from nine to 11 o'clock that evening, guests were invited to view 34 paintings, including "All Dressed Up for Sunday," "Bringing in the Hay," and "Apple Pickers." Moses stayed home. "We had invited the artist to the opening of the show but she declined," Kallir recalled, "saying she did not need to come since she knew all the pictures anyway."

Although Caldor played a critical role in discovering Moses' works, it was Kallir who believed in her talent and engineered her stunning success. One summer afternoon, I make my way to Galerie St. Etienne, where I meet Kallir's granddaughter, Jane Kallir, who is the gallery's owner and an authority on Moses' work. In her quiet office just off the gallery's exhibition space, she takes me back to the state of the art world at the time of Moses' debut.

On the one hand, says Jane, self-taught artists John Kane and Horace Pippin had been recently heralded as American iterations of Henri Rousseau, the French toll collector turned artist who had become a favorite of Picasso, Kandinsky, and other modern painters. At the same time, although some American champions of modern art supported untrained artists, others felt that the United States should cultivate a more "serious" form of modernism—an attitude that would deepen over the following years.

Otto Kallir, a Jewish-Austrian collector who fled the Nazis in Vienna, had an outsider's perspective. "My grandfather was highly unusual, perhaps because he wasn't American. Americans had this enormous chip on their shoulders. He didn't," she says. Unswayed by cultural prejudices, he was open to the country's native talents, including folk art and "primitive" paintings, as Moses' works were then known. "He had that ability to recognize quality in art and to understand that her use of color, her compositional inventiveness, her ability to grow and change as an artist, all of this was the real deal," says Jane. "He had a very good eye and a very clear sense of his own taste."

Jane herself wasn't quite sure what to make of the artist growing up, she tells me. Despite her grandfather's celebration of Moses' work, her American parents were far more typical in terms of their aesthetic biases. "They disdained Grandma Moses," she says. Of the two small Moses paintings they owned—given to them by Otto Kallir—one hung on top of a bookcase, where Jane could barely see it. As a result, she found herself far more interested in the Austrian artists her grandfather represented in his gallery, including Egon Schiele.

In college, Jane majored in studio art and art history. But in line with the formalist teaching of the 1970s, she learned nothing about Moses' work. It was only after graduating, when she began working at her grandfather's gallery, that she began to appreciate the artist for herself. "There was a Grandma Moses hanging, 'Grandma Goes to the Big City.' I remember just staring at that painting and looking at how she did it and all its component parts. I was just bowled over," she tells me. "And I said, 'This is a great artist.'"

It was her grandfather's endorsement of Moses' work that sent the artist's career into orbit. One month after her first show at Galerie St. Etienne, Gimbels department store exhibited her work as part of their 1940 Thanksgiving festival. This time, Moses agreed to come to the city. She arrived armed with a sampling of her homemade cakes and preserves and a chaperone, Carolyn Thomas, owner of Thomas's drugstore. With little warning, Moses found herself on stage in Gimbels' auditorium, where she and Thomas had a conversation in front of 400 people. "I was in from the back woods, and I didn't know what they were up to," she later wrote. "My, my, it was rush here, rush there, rush every other place—but I suppose I shouldn't say that, because those people did go to so much bother to make my visit pleasant."

Spark

Moses' persona appealed to the public. Reporters jumped on the story, describing the elderly artist as an indomitable, plucky little woman with a quick wit and a moniker to match: Grandma Moses, an honorific used by family and neighbors back home and adopted by everyone else. Jane Kallir, who met the artist when she was about five years old, remembers her warmly. "She went out of her way to be nice to me, to talk to me, to pay attention to me," she tells me. "She gave me a doll to play with. She was a real grandma, even if she wasn't my grandma."

With her wire-rimmed glasses, high-collared dresses, and silver hair piled in a bun on top of her head, Moses exuded old-world country charm. And she was authentic, making no pretense of who she was or where she most wanted to be. "I like the song of birds better than the honk of automobiles," she said, "and trees are prettier than skyscrapers."

Nature would always be Moses' world—the place she turned to for hope and continuity, and the subject matter of her paintings. "I like to paint old timey things, historical landmarks of long ago, bridges, mills, and hostelries, those old-time homes," she wrote. In contrast to plein air artists, who work outside with easels, Moses drew from experience, culling her mind for remnants of the past— the fields, trees, fences, animals, farmworkers, children, blossoms, and skies that filled her days in western Virginia and upstate New York. "Memory is history recorded in our brain," she wrote. "Memory is a painter, it paints pictures of the past and of the day."

For inspiration and guidance, Moses often foraged in a large trunk she filled with clippings from greeting cards, newspapers, and magazines, including *Life,* the *Saturday Evening Post,* and *National Geographic.* She especially liked prints by Currier and Ives, whose lithographs depicted iconic scenes of rural New England. Moses sometimes used the images she collected as templates or stencils, arranging them on her painting and tracing figures she liked. As a result, her paintings were often a blend of elements, both borrowed and her own.

Every kind of artwork, whether a cubist portrait or a country landscape, is powered by a painter's imagination. Moses harnessed hers in choosing her subjects and mapping out her compositions. This is critical to understanding her work, says Will Moses, the artist's great-grandson and a successful folk artist himself. Will and his wife live in his great-grandmother's old white farmhouse in Eagle Bridge, and he maintains his studio and gallery, Mount Nebo, in a red barn that sits just across the driveway. The view out front on a summer day is blissfully calm—corn fields, a medley of leafy trees, a jumble of clouds, a deep blue sky. A historical signpost announces the site as Moses' homestead and the artist as "World Renowned Painter of Rural Life."

Moses' son Forrest taught Will to paint when he was about five years old. By then, Grandma Moses was in her 90s and he and his cousins used to ride their bikes across the field to watch her work. "She would explain to us what she was doing, she'd give us molasses cookies and away we'd go," he recalls. "We all knew enough to pay attention."

Moses often said that she liked pretty pictures and bright colors and though she occasionally strayed into more dramatic terrain—a thunderstorm or a fire on a covered bridge—she stuck mostly to scenes that pleased her viewers. "They want to see how life was lived, but they don't want to see the dark side of it," Will tells me. "She was portraying that nostalgic life where times were simple. As long as chickens were laying eggs and cows were milking, things would be OK."

The artist's deep knowledge of her subject matter—she had lived it all her life—allowed her to control her images as she saw fit. "Often-times, her imagination improved upon the scenes considerably," says

Will. This was especially true of a painting Moses created of President Dwight Eisenhower's farm in Gettysburg, Pennsylvania, which his Cabinet commissioned to mark the third anniversary of the president's inauguration. Using black-and-white photographs as her guide, Moses rendered the scene faithfully, but she also took liberties. "Farmer Eisenhower is proud of his growing herd of black Angus cattle," the *New York Times* reported. "He owns only one Holstein and one brown Swiss. But Grandma Moses painted four Holsteins to three Angus cattle grazing in the pastures." She also expanded the size of his golf green, prompting the president to quip, "I wish it was that big."

Throughout her career, Moses stayed true to her roots. Unschooled in art and fully focused on her work, she had little interest in the great painters of the day and she did not concern herself with cultural trends gallivanting through the outside world. "She knows as little about Picasso's paintings as she does about Freud's writings," journalist S. J. Woolf wrote. "Impressionism and cubism have no meaning for her."

Just as Shirley Temple's exuberance cheered a nation desolated by the Depression, Moses' art offered refuge from a chaotic world: a panacea for the anxieties stirred by war and the tensions and insecurities that followed. "Her talent wasn't just composition or color," says Katherine Jentleson, curator of folk and self-taught art at the High Museum in Atlanta. "It was creating an atmosphere that people wanted to live in."

And it was a welcome sanctuary for Moses as well. "As I finish each picture, I think I've done my last," she once said, "but I go right on."

Wouldn't it be wonderful if we could peer into Moses' brain to see the neural machinations at work? What powered the painter's inventiveness, vigor, and stick-to-itiveness, allowing her to achieve such extraordinary heights in the eighth, ninth, and 10th decades of her life?

Science has long been interested in how our brains age and how malleable they are late in life. For much of the 20th century, scientific dogma held that brain "plasticity" was the purview of the young. Researchers believed that although the infant was capable of producing new brain cells after birth—a dynamic process known as neurogenesis—the adult brain was not.

But over the last decades this conviction has been challenged by reports in a variety of species, including birds, monkeys, and human beings, which suggest that the adult brain is far less rigid than previously believed. In one study, published in 2018, Columbia University researchers examined the brains of 28 human donors ages 14 to 79 who had died suddenly. Although the older brains were not as spry as their younger counterparts—they had lower levels of a protein involved in cell communication and fewer blood vessels, which supply oxygen—the scientists were surprised to find thousands of immature neurons in the hippocampus, an area involved in memory and learning. Neurogenesis in aging brains is an evolving field and there is still no absolute consensus. But the idea that grown adults are capable of forming new neurons is exceedingly promising news.

Longer life spans mean we have more time to make use of this brainpower. Since 1900, life expectancy has more than doubled worldwide, and although numbers fluctuate year to year (opioid overdoses contributed to a recent drop in the United States, and the coronavirus pandemic will have a global impact), the overall trend over the last 100 years is up: Humans are surviving much longer than we used to—on average into our 70s and in some cases, well into our 90s. These bonus years, once unimaginable, give us extra days, weeks, and months to experiment and take on new challenges, as long as we are able to maintain good health.

Spark

For some people, aging precipitates a desire for exploration. Consider Diana Nyad's 110-mile swim from Cuba to Florida at the age of 64 or Barbara Hillary's journeys to the North and South Poles in her 70s. Nyad emerged victorious in her fifth attempt to cross the Florida Straits after swimming for nearly 53 hours without a protective shark cage. Hillary, a cancer survivor, became the first Black woman on record to reach both ends of the earth. "At every phase in your life, look at your options," she said in an address to graduating seniors at the New School, her alma mater, in New York City. "Please, do not select boring ones."

Even as the body ages, the mind swaggers with vitality—at least among some inspirational souls. The average age of patent holders in the United States is 47, with as many inventors over the age of 50 as below the age of 40. Take University of Texas physicist John Goodenough. He filed his 27th patent when he was 94 years old and shared the Nobel Prize in 2019 at the age of 97 for developing the lithium-ion battery. "Some of us are turtles; we crawl and struggle along and we haven't maybe figured it out by the time we're 30," he said with abundant modesty to the *New York Times*. "But the turtles have to keep on walking."

When the mind is open to new experiences, creativity thrives, no matter one's age. British author Harry Bernstein saw his critically acclaimed memoir, *The Invisible Wall*, published when he was 96. Composer Elliott Carter wrote a new piece of music for his 100th birthday celebration at Carnegie Hall. Masako Wakamiya, a retired banker in Tokyo, developed Hinadan, an iPhone game app featuring traditional Japanese dolls, when she was in her 80s. "Curiosity makes me jump quickly to try new things," Wakamiya said in an interview. "I don't make walls to shut out unknown worlds."

It's absolutely possible to accept the reality of aging and also relish its gifts. Few have done this better than the neurologist and writer

Grandma Moses

Oliver Sacks. In an essay written on the cusp of his 80th birthday in 2013, Sacks wrote that his father deemed his 80s one of his most enjoyable decades. "He felt, as I begin to feel, not a shrinking but an enlargement of mental life and perspective," wrote Sacks. "I do not think of old age as an ever grimmer time that one must somehow endure and make the best of, but as a time of leisure and freedom, freed from the factitious urgencies of earlier days, free to explore whatever I wish, and to bind the thoughts and feelings of a lifetime together."

With scientific merriment, Sacks marked his years in atomic numbers. "At eleven, I could say, 'I am sodium' (Element 11)," he wrote in his essay. Soon after the piece was published, I went to see Sacks interviewed at Sixth & I, a synagogue and arts space in Washington, D.C., where he talked about his latest book, *Hallucinations*—one of six published after he turned 65. Sure enough, he appeared on stage in a bright blue T-shirt emblazoned with "Hg," the symbol for mercury (Element 80). Although he was a bit unsteady and used a cane, his joy seemed invincible. I remember wishing I could leap from my balcony seat and catch it in a butterfly net.

What do we make of creative powers that emerge unexpectedly late in life? In his book, *Musicophilia,* Sacks explored the curious onset of musical and artistic talent in some people who suffer damage to regions of the brain that control inhibition following a stroke—a scenario reminiscent of savants who exhibit remarkable dexterity in certain subjects despite suffering brain injuries early in life. Sacks wondered if the brains of healthy people who discover late-life abilities may be functioning in a similar way—by releasing and unlocking their own mental handcuffs, they allow creativity to flow, an effect Sacks dubbed "the Grandma Moses phenomenon."

Interestingly, creativity sometimes improves with age among artists, with some doing their most notable and recognized work in their

20s and 30s and others peaking much later—in their 50s and beyond. In his book, *Old Masters and Young Geniuses,* University of Chicago economics professor David Galenson reported that age can enhance the value of an artwork. For Picasso, the prodigy, bidders paid the highest prices at auction for works he did at age 26; Cézanne, on the other hand, had his best year when he was 67. Despite a cultural infatuation with early achievement, "there will always be the Cézannes of the world," Galenson said in an interview with *Smithsonian* magazine, "though we may not know who they are until they are in their 60s or 70s or 80s."

All of this raises the question: Did Grandma Moses *need* to be old to do her best work? If she'd had the opportunity to start earlier in life, would it have made a difference? Clearly, she would have had more time to develop as an artist, to deepen her skills and experiment with new techniques. At the same time, Moses' age played a critical role in the way she approached her craft and in the content she chose to paint. As a late-in-life artist, her works revealed what she treasured most about her life.

Many people attempt to make sense of their later years by reminiscing about the past. Moses was likely doing the same as a storyteller, both in words and images, Jane Kallir observes. "I think there is a retrospective quality to what she is doing," she says. Although Moses suffered great losses in the premature deaths of some of her siblings and children, she presented "a perspective of contentment and gratitude," says Leslie Umberger, curator of folk and self-taught art at the Smithsonian American Art Museum. "She was proud of who she was and the life she had lived."

On a visit to the Bennington Museum in Vermont, which holds the world's largest public collection of Grandma Moses' work, I notice that although the artist depicted many of the same subjects repeatedly—an old checkered house, the rural farm landscape, people bundled up tapping maple trees for sap—her paintings seemed to connect, one to the next, creating a panorama of snapshots. Umberger confirmed to me the importance of viewing her work in its entirety. "Collectively her paintings reveal the arc of a life, functioning together like the pages of a book," she says. "Their impact is strongest when a number of pieces are seen together."

In the museum, I notice a woman standing quietly in front of one of Moses' paintings. She tells me she is "80-plus" and deeply moved by the artist's work. "I love Grandma Moses. I love it because it brings me back—I'm going to cry—to a happier day," she says. "There's so much bad stuff that goes on today. It makes you so happy. It's a gift from God."

One of the greatest perils of aging is losing one's sense of purpose. Our daily activities, whether going to school or showing up for our jobs, shape our identities and give us direction. Moses started working at the age of 12, and she wasn't about to lie down on the couch and die, says her great-grandson Will. "She wanted to keep going, she wanted to feel productive."

Feeling useful is good for both body and mind. Studies show that people who have a sense of purpose in life take better care of their health and are at lower risk for disease. Researchers at the University of Michigan recently studied nearly 7,000 people over the age of 50 for four years and found that those who had the highest "life purpose" were less likely to die from heart, circulatory, and blood conditions. Those who did not have a strong purpose in their lives, meanwhile, were more than twice as likely to die before the study's end.

Moses didn't need studies to know that keeping busy was good for the soul. But she had another equally important reason to make art:

Spark

She needed income. As her paintings began to sell, her work became a sensible undertaking. "To some degree, I think she saw painting as being not that different from baking pies," says Will. "She was providing for her family." Moses stated this plainly herself: "If I didn't start painting, I would have raised chickens."

Over the course of 21 years as an artist, Moses watched her paintings travel to Vienna, Munich, Salzburg, Berne, The Hague, and Paris. And she saw their price tags soar—from just a few dollars at Thomas's drugstore to as much as $10,000 during her lifetime. She got to know Lillian Gish, who portrayed Moses in a television drama; cut her 89th birthday cake with her friend Norman Rockwell, who lived in nearby Vermont; painted a picture in real time under television lights for Murrow's first colorized edition of his popular show, *See It Now;* appeared on the covers of *Life* and *Time* magazines; and sold millions of Hallmark Christmas cards featuring her snowy winter scenes.

Throughout it all, her vigor defied her years. In 1949, when she was 88, Moses joined five other women, including Eleanor Roosevelt, to receive an award from the Women's National Press Club in Washington, D.C. In a "My Day" column that followed, the former first lady, then in her mid-60s, took note of Moses' stamina. "She stood and shook hands with several hundred people," Roosevelt wrote, "and wouldn't even listen to a suggestion that she had to have special care."

After World War II, as abstract expressionism became America's celebrated art form, folk artists struggled to compete with the likes of Jackson Pollock, Willem de Kooning, and Mark Rothko. Still, Moses never lost her extraordinarily popular appeal. Even as many in the mainstream art world viewed Moses and her work as quaint or passé, says Umberger, she was still "the most famous female painter of her day." An exhibition of 43 of her paintings at the National Gallery of Art in Washington, D.C., in 1979 drew 137,000 viewers—more than the 98,056 who went to see a small display of Piet

Mondrian that same year and almost as many as the 148,415 who flocked to view a traveling exhibition of the Hermitage's collection of Italian Renaissance painters, which included Correggio, Leonardo, and Titian.

Six decades after her death, Moses' place in American art history is gaining renewed attention, and her work has enjoyed something of a revival. "The paintings are always engaging, sometimes marvelous, and, after a long rest in genteel approbation, as good as new," *New Yorker* critic Peter Schjeldahl wrote in 2001 about an exhibit Jane Kallir curated at the National Museum of Women in the Arts. In 2006, Moses' painting, "Sugaring Off," sold for $1.4 million; in 2017, a bidding war quintupled Sotheby's estimated price for another one of her works, "Sugaring Time," made with oil and glitter on Masonite, which landed a final sale of $519,000.

The biggest boost for the artist is likely still to come. In 2018, the Smithsonian American Art Museum announced plans for a major traveling exhibition of Moses' work, kicking off in Washington, D.C. The hope is to highlight Moses' most iconic paintings, says Umberger, and position her among other American artists as "one of the most significant painters in our country's history."

By the time Will Moses got to know his great-grandmother, she had moved her painting supplies into the laundry room. Her family had built her a studio and a sun porch in the new one-level home she had moved into across from the old white farmhouse. But she wanted an escape from the fans and well-wishers who trooped out to Eagle Bridge to meet her, Will tells me. While family members provided cover, Moses holed up in her little workroom, making art in the company of her washer and dryer.

Although the attention could be overwhelming and exhausting at times, Moses was well aware of what a noteworthy life she had lived. Over a century, she had farmed fields, raised children, witnessed wars, taught herself to paint, traveled to the big city, met political leaders, and achieved fame on an international scale. "Here was this woman from Washington County who was making a name for herself because of what she was doing with paint," says Will. "I think she thought that was pretty cool."

On September 7, 1960, Moses' 100th birthday, letters and tributes poured in to Eagle Bridge from around the world. President Eisenhower wrote that "all of us benefit from your sense of the joy and beauty of life." The director of the Musée National d'Art Moderne in Paris said Moses "makes us realize that there is still a bit of paradise left on this earth." And the *London Art News and Review* hailed Moses as "one of the key symbols of our times. A real artist whose paintings reveal a quality identical with genius."

One year later, in the summer of 1961, Moses was admitted to the Hoosick Falls Health Center after a fall. She wasn't happy. When her longtime doctor, Clayton E. Shaw left his stethoscope in her room, she hid it and told him she wouldn't return it until he took her back to her home in Eagle Bridge. As the date of her 101st birthday approached, Shaw called off all celebratory festivities, saying that the artist needed to take it easy. "She has begged her family to bring her paints to the Health Center," the *Times Record* reported, "but at the suggestion of Dr. Shaw they have put it off. If she has her paints, she won't rest, the doctor says."

Moses died peacefully on December 13, 1961. The cause was hardening of the arteries, Shaw said, but also added, "she just wore out." Albany Congressman Leo W. O'Brien praised the artist as having given "more encouragement to old people than almost anyone of our time. She was a living example that talent didn't end at 40 years of age."At 101, Moses outlived three of her five children and left behind two sons, one daughter-in-law, 11 grandchildren, and 30 great-grandchildren.

At the end of his interview with Moses, Murrow asked the artist about her plans for the next 20 years. "I'm going up yonder," she said, pointing her brush to the heavens with a smile. "Naturally, naturally, I should. After you get to be about so old, you can't expect to go on much farther."

So you have no fear or apprehension? Murrow asked. "Oh no, no, you go to sleep and wake up in the next world," the artist responded. "Well," said Murrow, "you'll leave behind more than most of us will."

"I don't know about that, I don't know about that," Moses said.

Then she looked down at the painting on the table in front of her and reached for a brush.

Leonardo da Vinci

Eternal Genius

G enius lures me in.

I know from reading hundreds of books and articles and interviewing countless experts that the subject has a complicated and, at times, distasteful past—most notably when researchers focused solely on bloodlines, elevating one lineage over another, or used IQ as the sole yardstick for potential.

Women have long been overlooked and it is our collective loss when their stories are not more widely celebrated. We all learn about Shakespeare, but how many know Murasaki Shikibu? The Japanese writer, born around the year 978, wrote *The Tale of Genji,* which scholars consider a novelistic tour de force. The contributions of x-ray crystallographer Rosalind Franklin are subsumed by Watson and Crick. And until the movie *Hidden Figures* came out, few had heard

Spark

of the astute NASA mathematicians Dorothy Vaughan, Mary Jackson, Katherine Johnson, and Christine Darden. Above all, brilliance is far too often celebrated among those born into privilege, while neglected in those who never get the chance to be nurtured.

Still, the exploration of genius demystifies its key elements—intelligence (of all kinds), creativity, persistence, and luck—and allows us a glimpse into our own potential. I, for one, have learned how to cultivate imagination, push through writing blocks, and listen to the muse—even if I cannot tell you exactly who or what she is.

When the ingredients of genius come together, as they have in the 12 individuals that grace these pages, the result is often mystifying and dazzling at once. How do some people overcome misfortune to become innovators in music, business, and science? What are the moments of providence that galvanize achievement? Why does genius ignite in one person soon after birth and in the twilight of life for another? These are the inquiries that drove this undertaking.

My own spark—the moment my curiosity erupted and the questions began to cascade—rests with Leonardo da Vinci, the Renaissance man who embodied genius in the voracious way he puzzled out the world. I became a student of Leonardo before I began researching the figures in this book; while writing about them, I couldn't stop thinking about him.

The artist seems a fitting coda to this journey, not only because of his unparalleled ability to observe, interrogate, and visualize, but also because of the strands of genius that connect him to the individuals in this book—and to the rest of us, the beneficiaries of his work.

My journey into the life and mind of Leonardo began with a visit to the Italian village of Vinci, where Leonardo was born on April 15, 1452, then looped through the places he lived and the works he left behind. I interviewed scholars who have dedicated their lives to studying the artist's work. I read translations of Leonardo's manuscripts and viewed as many of his paintings as I could, including "Ginevra de' Benci," "Adoration of the Magi," "Mona Lisa," and "Salvator Mundi." I traced the artist's footsteps in Florence and visited du Clos Lucé, the modest chateau in France's majestic Loire Valley where Leonardo spent his final years before his death on May 2, 1519.

It was not until I had the privilege of viewing a selection of Leonardo's drawings, preserved by the Royal Collection Trust at Windsor Castle, that the scope of his genius soared into view. Seeing the depth of the artist's intellectual wanderings in his own hand was staggering. In ink, chalk, and silverpoint, he hunted for detail while dashing from one subject to the next: botany, biology, geology, hydraulics, architecture, astronomy, military engineering, costume design, geometry, cartography, optics, and anatomy. Leonardo sketched to make sense of the universe.

The images are breathtaking in their lucidity. The most diminutive, a fragment smaller than a thumb, shows a female torso depicted in just a few muted strokes. The most iconic, rendered tenderly in red chalk and curved hatch marks, depicts a fetus curled up in a womb. Everything is put to the test with visual precision: a study of drapery on Madonna's arm; mortars bombarding a fortress; the umbra and penumbra of shadow; a skull, a heart, a foot, and the sweep of the human face—from the luminescence of Leda to the haggard expression of an elderly man.

Texture and blemishes pop from the pages, and one can see the vigor with which Leonardo worked—the intensity of color he explored in his paper washes and chalk, the energy of his brush marks, and his remarkable dexterity. "What you get most out of Leonardo's drawings is this completely unfettered way of leaping between subject matter," Martin Clayton, head of prints and drawings for the Royal Collection Trust, tells me. "There's something tremendously exciting about seeing a mind work in this incredibly broad way."

Spark

⌐⌐

Leonardo's boundless genius emerged from humble origins. Born out of wedlock to Ser Piero da Vinci and a woman believed to be Caterina di Meo Lippi, a local peasant, Leonardo spent his early childhood in and around Vinci, dotted with olive groves and dusky hills in the Tuscan landscape.

At some point in the early part of his life, Leonardo began sketching; by his teenage years he had landed an apprenticeship in Florence with artist Andrea del Verrocchio, with whom he collaborated on religious paintings and on the copper ball that sits atop Brunelleschi's dome. Leonardo's artistry soon upstaged not only his peers, but also his mentor. Within several years, he'd received his first commissions: an altarpiece for a chapel in the Palazzo della Signoria and "Adoration of the Magi," depicted in charcoal, watercolor ink, and oil on wood.

Leonardo was lucky to live at a time when his abilities and interests coincided with a passion for his work. During the Italian Renaissance, stretching roughly between the 14th and 17th centuries, wealthy patrons cultivated the arts. Inventiveness coursed through the streets. Great artists of this era included Michelangelo, Raphael, Botticelli, Titian, and Tintoretto, as well as female painters who are far less well known: Marietta Robusti (Tintoretto's daughter), Plautilla Nelli, Sofonisba Anguissola, and Lucrezia Scarfaglia.

Although Leonardo left few personal reminiscences of his own, we have glimmers of the man. He was almost certainly gay—his lifelong companions were male and he was twice accused of sodomy, though charges were dropped in both cases. An animal lover, he bought caged birds at market and set them free. He wore rose-colored tunics and was admired for his singing voice, generosity of spirit, and his social finesse. He would have been a very entertaining dinner guest, says Gary Radke, emeritus professor of art history at Syracuse

University. "He wasn't one of these inscrutable, pondering, grousing geniuses."

Throughout his 46-year career, spent largely in Florence and Milan, Leonardo willed himself to knowledge, touched by an ever wandering eye and the determination to follow it. He learned Latin, collected poetry, and read Euclid and Archimedes. Where others embraced the perceptible, he scrutinized minutiae—geometric angles, the dilation of the pupil—bounding from one discipline to the next while seeking links between them. He sketched flowers and flying machines; designed war machines for his patron, Duke Ludovico Sforza; crafted theatrical ornaments out of peacock feathers; and engineered a plan to divert the Arno River between Florence and Pisa.

The pursuit of knowledge never stopped. Leonardo's to-do lists included jottings to "construct glasses to see the moon larger" and "describe the cause of laughter." He sought answers to an endless supply of questions: What's the distance from the eyebrow to the junction of the lip and the chin? Why are stars visible by night and not by day? How do the branches of a tree compare with the thickness of its trunk? What separates water from air? Where is the soul? What are sneezing, yawning, hunger, thirst, and lust?

Leonardo documented everything in magnificent precision on the backs and corners of paper with tidy notes written with his left hand in mirror script, from right to left; nobody quite knows why. Some of these pages exist as loose sheets today; others have been bound into the volumes now known as notebooks or codices. There's no clear order, even on a single page, and similar themes often appear on different sheets completed years apart.

All of this makes it challenging even for scholars to keep up with the brisk tempo of his mind, says Paolo Galluzzi, director of the Museo Galileo in Florence. I meet Galluzzi in his office at the museum one day, where he thumbs through reproductions of Leonardo's notebooks

Spark

with a sense of wonder. Every time the artist made an observation, a question arose in his mind, which invariably led to another, he tells me. "He went sideways."

It is difficult to grasp Leonardo's unparalleled ability to push past the work of his forebears. He did this by cross-examining his subjects and overturning his own verdicts. In the Codex Leicester, owned by Bill Gates, Leonardo tirelessly investigates how water makes its way to mountaintops, ultimately rejecting his initial conviction that heat draws it upward. Instead, he realized, water circulates through evaporation, clouds, and rain. "More important than discovering how mountain streams work was discovering how you would discover it," says his biographer Walter Isaacson. "He helps invent the scientific method."

For Leonardo, the precepts of science—observation, experiment, and experience—were critical to his art. He moved fluidly between the two realms, grasping lessons from one to inform the other, says Francesca Fiorani, associate dean for the arts and humanities at the University of Virginia. One of his greatest gifts was his ability to make knowledge visible, she says. "That's where his power is."

Nowhere is this more vivid than in Leonardo's study of anatomy. He dissected human cadavers, teasing out underlying musculature in three dimensions to sort out how a leg bends or an arm cradles. Leonardo's contemporaries, including his rival Michelangelo, studied muscles and bones to improve their artistic representation of the human body. "But Leonardo went beyond this," says Domenico Laurenza, a science historian based in Rome. "His approach to anatomy was that of a real anatomist."

The scientific data Leonardo collected in his notebooks underlie every stroke of his paintbrush. His analysis of light and shadow allowed him to illuminate contours with unparalleled subtlety. He did away with traditional outlining, instead softening the edges of figures and objects in a technique known as sfumato. Optics and geometry led to a sophisticated sense of perspective, exemplified in "The Last Supper," painted in the 1490s. His anatomical studies drilled down on the biology of facial expressions. Which nerve, he queried in his notes, causes "frowning the brows" or "pouting with the lips, of smiling, of astonishment." His observations allowed him to depict emotional depth in the people he painted, who appear sentient rather than stiff.

Leonardo's inventiveness came with a price. He irked his patrons with incessant delays, and many of his works went unfinished, including the "Adoration of the Magi" and "The Virgin and Child with Saint Anne." Scholars have attributed these propensities to his exuberance for new subjects and his perfectionism. It was also because the challenge of doing outweighed the expectation of getting it done. For Leonardo, it's all about process, says Carmen Bambach, curator of drawings and prints at the Metropolitan Museum of Art. "It's not really about the endgame."

Indeed, the more knowledge Leonardo acquired through the studies in his notebooks, the more difficult it became to see a finish line in his art. "As he kept painting, he understood that you could create such infinitesimal gradations of tone and transition from the highest, most intense highlight to the deepest shadow," says Bambach. Analyses of Leonardo's work reveal copious revisions, known as pentimenti. Infinity became a very real concept that took on practical implications: There was always more to learn. "In many ways, intellectually, this is an unending process," she says.

This may help explain why Leonardo never published his notebooks. He intended to complete treatises on numerous subjects, including geology and anatomy. Instead, his sketches and manuscripts were left to his faithful companion Francesco Melzi to sort through. In the decades following Leonardo's death, two-thirds to three-quarters of his original pages were likely pilfered or lost, making the sweeping reach of his productivity all the more astounding. It was not until the late 18th century that most of the surviving pages began to be published, more than 200 years after he died. Centuries later, we are still catching up with him.

⌣

There will never be another Leonardo da Vinci. But the most important revelation I have gleaned from his work is this: There will forever be an abundance of inventive ideas and individuals who pursue them. Genius is far more complex than any of us can comprehend—but strands of genius will always exist, manifest in distinct ways in different people.

As I pieced together the trials and triumphs of the individuals in this book, I discovered clear echoes of the artist filtering through the lives of those who came after him. Like Leonardo, who kept a small book of pages tied to his belt, Isaac Newton took copious notes—at one point in what he called a "waste book," in which he detailed his mathematical and optical calculations. At a much later date—and in a completely different realm—Sara Blakely has filled her shelves with the notebooks she has carried with her for more than two decades. Ideas are precious, she says. "The minute we get them we should write them down."

Leonardo's constant quest for knowledge through experience—testing the turbulence of water and the flow of air—speaks to the inexhaustible churning of his mind. "Just as iron rusts from disuse, and stagnant water putrefies, or when cold turns to ice, so our intellect wastes unless it is kept in use," he wrote.

Roget achieved this end by refusing to retire after a successful medical career; instead, he turned obsessively to words. Eleanor Roosevelt pushed herself to learn from people and to take action—to go down into the coal mines—rather than observe from the sidelines. "*You must be interested in anything that comes your way,*" she emphasized in her book, *You Learn by Living.* Bill Gates keeps his tote bag of books by his side. "I read and learn new things and talk to experts as much as I can," he tells me. Among all of Leonardo's vast achievements, Gates writes, "the attribute that stands out above all else was his sense of wonder and curiosity."

And what of leaps and bounds and breaking barriers? Leonardo visualized contraptions that were yet to be invented: diving suits and flying machines. Picasso followed in the way he reshaped the human face and figure, forcing us to ask questions about how we perceive what we think we know. Maya Angelou trampled convention in her writing, employing a rawness and candor that allow us to contemplate not only one woman's life, but also an entire history of people: the unsparing experience of Blacks in America, post-slavery.

Leonardo's inquiries, which roamed without borders, reverberate in the way Yo-Yo Ma stretches across musical genres, uniting the strings of his centuries-old cello with the rhythms of bluegrass fiddlers, African drummers, and a Galician bagpiper to create new sounds. The edge where two entities meet makes the center stronger, Ma says. "It's the place where a lot of creativity happens."

Focusing on the smallest elements to understand the whole was Leonardo's forte; he bore down on what was in front of him. Alexander Fleming discovered penicillin, the antibiotic that revolutionized medicine and saved millions of lives, thanks to his keen eye; he patiently scrutinized what another scientist may have disregarded. Serendipity made a cameo, but Fleming starred. "It was not an accident that made him great," the *New York Times* noted in his obituary, "but his training, his knowledge, his obstinate, inquiring mind."

More than anything else, Leonardo stood for pushing the boundaries of learning. Like him, every one of the figures in these pages was self-taught in one way or another. As a toddler, Shirley Temple figured out where to stand on stage by interpreting the heat from a spotlight on her face. Julia Child studied at the Cordon Bleu, but her recipes evolved from the trials of adding too much salt or bungling her potato pancake flip. Anna Mary Robertson Moses figured out how to paint the swirling clouds and sledding children that dot her landscapes by picking up her brush and dipping it into a jar of paint. For Moses, the conceptions painters came up with themselves mattered most. "If they have a teacher," she said, "they will soon paint as the teacher paints. And it's best for them to use their own ideas."

Perhaps the most defining characteristic of genius is the impact on future generations. Great minds reinvent and reconceptualize tradition, leaving the rest of us in a new space that offers unseen potential. This became tangible in the most exquisite way for me and my family one summer evening in 2018, as we watched Polish pianist Sławomir Zubrzycki stride into an expansive hall at Kalmar Castle on the southern coast of Sweden. Dressed in a formal waistcoat and polished black shoes, Zubrzycki sat down to play a concert of Renaissance music on an instrument envisioned—but never built—by Leonardo da Vinci.

Among his many pursuits, Leonardo improvised melodies on the lira da braccio, a Renaissance-era string instrument, and parsed the intricacies of acoustics and musical design in his notebooks. In 2009, Zubrzycki found himself transfixed by a series of the artist's sketches for a viola organista, a keyboard instrument with bowed strings. Captivated by the possibility of one instrument fusing two musical families, Zubrzycki set out to build it.

None of Leonardo's drawings offered a detailed blueprint for Zubrzycki to follow. For four years, he spent five hours a day researching and formulating his design. He tested wood samples, resolved that he needed 61 keys, and puzzled out how to build four circular bows covered in horsehair that could rub against a series of strings to create a melodious sound. As he brought the instrument to life, Zubrzycki drew on the same vital force that drove Leonardo: his imagination.

Although Zubrzycki's viola organista looks like a baby grand piano, it performs like a full-bodied string ensemble—a singular experience for listeners. As Zubrzycki played, a fusion of stringed notes filled the hall as if performed by groups of musicians on multiple instruments. Resounding and joyful, the rich and complex sound of Zubrzycki's viola organista evoked the luminescence of Leonardo's paintings: a musical sfumato with soft edges and resonant tones. Like Leonardo's art, the melodies lingered.

Leonardo ranked music as second only to painting—a higher pursuit, even, than sculpture, describing it as *"figurazione delle cose invisibili*—the shaping of the invisible." For Zubrzycki's audience, such an exalted moment occurred in a castle as the sun set over the Baltic Sea, when a few scribbles in Leonardo's notebooks morphed into music.

The quest for infinite knowledge is ours to grasp. Numbers, years, decades—these are signposts, not rules over the fertile provinces of our minds. I conclude with the words that scholar Paolo Galluzzi said about Leonardo, which have long stuck with me: "He went sideways."

Go sideways. See where the spark will take you.

Sources and Notes

Throughout the course of my research, I conducted interviews with academic scholars, biographers, neuroscientists, and psychologists. A wide variety of additional sources informed my understanding of the mind, human behavior, and genius—and, more specifically, the science of prodigies, midlifers, and late bloomers. Because details from this research are seeded throughout the book, I am grouping these overview materials below. Source material for individual chapters follows.

SCIENCE OF GENIUS

Mihaly Csikszentmihalyi, *Creativity: The Psychology of Discovery and Invention* (New York: Harper Perennial Modern Classics, 2013); Angela Duckworth, *Grit: The Power of Passion and Perseverance* (New York: Scribner, 2016); David Epstein, *Range: Why Generalists Triumph in a Specialized World* (New York: Riverhead Books, 2019); Anders Ericsson and Robert Pool, *Peak: Secrets From the New Science of Expertise* (New York: Mariner Books, 2017); Howard Gardner, *Extraordinary Minds: Portraits of 4 Exceptional Individuals and an Examination of Our Own Extraordinariness* (New York: Basic Books, 1997); William B. Irvine, *Aha! The Moments of Insight That Shape Our World* (New York: Oxford University Press, 2015); Kay Redfield Jamison, *Exuberance: The Passion for Life* (New York:

Spark

Vintage Books, 2005); Claudia Kalb, *Andy Warhol Was a Hoarder: Inside the Minds of History's Great Personalities* (Washington, DC: National Geographic Society, 2016); Scott Barry Kaufman and Carolyn Gregoire, *Wired to Create: Unraveling the Mysteries of the Creative Mind* (New York: Tarcher Perigee, 2016); Mario Livio, *Why: What Makes Us Curious* (New York: Simon & Schuster, 2017); Darrin M. McMahon, *Divine Fury: A History of Genius* (New York: Basic Books, 2013); Joel N. Shurkin, *Terman's Kids: The Groundbreaking Study of How the Gifted Grow Up* (Boston, MA: Little, Brown and Co., 1992); Dean Keith Simonton, *Genius 101* (New York: Springer Publishing, 2009); Simonton, *Greatness: Who Makes History and Why* (New York: Guilford Press, 1994); Simonton, ed., *The Wiley Handbook of Genius* (Chichester, UK: John Wiley & Sons, 2014).

PRODIGIES, MIDLIFERS, LATE BLOOMERS

Patricia Cohen, *In Our Prime: The Fascinating History and Promising Future of Middle Age* (New York: Scribner, 2012); David W. Galenson, *Old Masters and Young Geniuses: The Two Life Cycles of Artistic Creativity* (Princeton, NJ: Princeton University Press, 2007); Barbara Bradley Hagerty, *Life Reimagined: The Science, Art, and Opportunity of Midlife* (New York: Riverhead Books, 2016); Ann Hulbert, *Off the Charts: The Hidden Lives and Lessons of American Child Prodigies* (New York: Knopf, 2018); Rich Karlgaard, *Late Bloomers: The Power of Patience in a World Obsessed With Early Achievement* (New York: Currency, 2019); Jonathan Rauch, *The Happiness Curve: Why Life Gets Better After 50* (New York: Thomas Dunne Books, 2018); Gail Sheehy, *Passages: Predictable Crises of Adult Life* (New York: Ballantine Books, 2006); Barbara Strauch, *The Secret Life of the Grown-Up Brain: The Surprising Talents of the Middle-Aged Mind* (New York: Penguin Books, 2010).

In my research on the elements of genius, I also consulted hundreds of articles in the popular press and scientific journals. Although too numerous to cite, I direct readers to these online sources, which provide stories and programs that may be of interest:

Sources and Notes

brainpickings.org; nobelprize.org; pbs.org; royalsociety.org; sciencefriday.com; scientificamerican.com.

～～

I consulted copious sources to inform my biographical research on the 13 individuals I profile, including: autobiographies, biographies, diaries, letters, magazine articles, newspaper stories, documentaries, and radio and television interviews. In my notes below, I have highlighted select materials mentioned within each chapter as well as some additional sources that proved especially informative.

INTRODUCTION
The introduction draws from material published in a cover story I wrote for *National Geographic:* "What Makes a Genius," published in May 2017.

CHAPTER 1: Pablo Picasso
This chapter draws from material published in a cover story I wrote for *National Geographic:* "How Picasso's Journey From Prodigy to Icon Revealed a Genius," published in May 2018.

Books and Exhibition Catalogs
Alfred A. Barr, ed., *Picasso: Forty Years of His Life* (New York: Museum of Modern Art, 1939); Gilberte Brassaï, *Conversations With Picasso* (Chicago: University of Chicago Press, 2002); Jennifer Drake and Ellen Winner, "Why Deliberate Practice Is Not Enough: Evidence of Talent in Drawing," in *The Science of Expertise: Behavioral, Neural, and Genetic Approaches to Complex Skill*, edited by David Z. Hambrick, Guillermo Campitelli, and Brooke N. Macnamara (New York: Routledge, 2018); Françoise Gilot and Carlton Lake, *Life With Picasso* (London: Virago Press, 1990); Patrick O'Brian, *Picasso: A Biography* (New York: W.W. Norton, 1994); Fernande Olivier, *Loving Picasso: The Private Journal of Fernande Olivier*, translated

Spark

by Christine Baker and Michael Raeburn (New York: Harry N. Abrams, 2001); Roland Penrose, *Picasso: His Life and Work* (New York: Schocken Books, 1962); Marina Picasso, *Picasso: My Grandfather* (New York: Riverhead Books, 2001); Olivier Widmaier Picasso, *Picasso: An Intimate Portrait* (London: Tate Publishing, 2018); Widmaier Picasso, *Picasso: The Real Family Story* (New York: Prestel Publishing, 2004); John Richardson, *A Life of Picasso: The Prodigy, 1881-1906* (New York: Alfred A. Knopf, 2007); Richardson, *A Life of Picasso: The Cubist Rebel, 1907-1916* (Alfred A. Knopf, 2007); Richardson, *A Life of Picasso: The Triumphant Years, 1917-1932* (Alfred A. Knopf, 2007); Theodore Roosevelt, *History as Literature and Other Essays*, (London: John Murray, 1914)*; Leo Stein, "More Adventures" in *Four Americans in Paris: The Collections of Gertrude Stein and Her Family* (New York: Museum of Modern Art, 1970); Antonina Vallentin, *Picasso* (Garden City, NY: Doubleday & Company, 1963); Ellen Winner, "Child Prodigies and Adult Genius: A Weak Link," in *Handbook of Genius,* edited by Dean Keith Simonton (New York: Wiley-Blackwell, 2014).

Newspapers, Magazines, and Scientific Journals

Associated Press, "Picasso Paid Tribute with Louvre Exhibit," in *Oakland Tribune,* October 24, 1971; Associated Press, "Pablo Picasso, Giant of Painting, Dies at 91," in *Fort Worth Star-Telegram,* April 9, 1973; Jennifer Drake and Ellen Winner, "Children Gifted in Drawing: The Incidence of Precocious Realism," *Gifted Education International* 29, No. 2 (2012), 125-139; Jennifer Drake and Ellen Winner, "Who Will Become a 'Super Artist'?" *The Psychologist* 26, No. 10 (October 2013), 730-733; Faraz Farzin et al., "Piecing It Together: Infants' Neural Responses to Face and Object Structure," *Journal of Vision* 12, No. 6 (December 2012); André Malraux, "As Picasso Said, Why Assume That to Look Is to See?" *New York Times,* November 2, 1975; Adrian Searle, "Storms of Colour From a Wild Destructive Genius," *Guardian,* May 29, 2019; Edward Vessel et al., "The Brain on Art: Intense Aesthetic Experience Activates the Default Mode Network," *Frontiers in Human Neuroscience* 6 (April 2012); Ellen Winner, "Exceptional Artistic Development: The Role of Visual Thinking," *Journal of Aesthetic Educa-*

tion 27, No. 4 (Winter 1993), 31-44; Ellen Winner, "How Can Chinese Children Draw So Well?" *The Journal of Aesthetic Education* 23, No. 1 (Spring 1989), 41-63; Yi Zhou, "Narcissism and the Art Market Performance," *The European Journal of Finance*, Vol. 23, No. 13 (2017), 1197-1218.

Additional Resources

Rodri García, "El Ayuntamiento publica el facsímil de la crítica de Picasso," February 14, 2015, *lavozdegalicia.es/noticia/coruna/coruna/2015/02/14/ayunta miento-publica-facsimil-critica-picasso/0003_201502H14C5992.htm; gagosian. com; museupicasso.bcn.cat/en; museopicassomalaga.org/en; museepicassoparis.fr/en; metmuseum.org; tate.org.uk.*

CHAPTER 2: Shirley Temple

Books

Shirley Temple Black, *Child Star: An Autobiography* (New York: McGraw Hill, 1988); Gerald Clarke, *Get Happy: The Life of Judy Garland* (New York: Random House, 2000)*; Anne Edwards, *Shirley Temple: American Princess* (New York: William Morrow and Co., 1988); Norman Eisen, *The Last Palace: Europe's Turbulent Century in Five Lives and One Legendary House* (New York: Crown, 2018); Kristen Hatch, *Shirley Temple and the Performance of Girlhood* (New Brunswick, NJ: Rutgers University Press, 2015); John F. Kasson, *The Little Girl Who Fought the Great Depression: Shirley Temple and 1930s America* (New York: W.W. Norton & Co., 2014); Amy Wallace, *The Prodigy: A Biography of William James Sidis, America's Greatest Child Prodigy* (New York: Dutton, 1986)*; Robert Windeler, *The Films of Shirley Temple* (Secaucus, NJ: Citadel Press, 1978).

Newspapers, Magazines, and Scientific Journals

Joan Acocella, "Not a Pink Toy," *New Yorker,* March 18, 2014; BBC News, "Obituary: Shirley Temple," February 11, 2014; Shirley Temple Black, "Prague Diary," *McCall's,* January 1969; Black, "Don't Sit Home and Be Afraid," *McCall's,* February 1973; Gerald Caplan, "An Outpouring of Love for Shirley Temple

Spark

Black," *McCall's*, March 1973; Marian Christy, "Shirley Temple at Age 53," *Boston Globe*, June 5, 1983; Douglas W. Churchill, "Hollywood: Tourists Accommodated," *New York Times*, March 22, 1936; Lawrence E. Davies, "Shirley Temple to Run for House From California," *New York Times*, August 30, 1967; Mordaunt Hall, "Shirley Temple, Adolphe Menjou and Dorothy Dell in the New Film at the Paramount," *New York Times*, May 19, 1934; Aljean Harmetz, "What Makes Shirley Run, in Her Own Words," *New York Times*, October 25, 1988; C. Robert Jennings, "Shirley Temple: Her Eyes Are Still Dancing," *Saturday Evening Post*, June 5, 1965; Robert Lindsey, "A Ruby Slipper for Shirley Temple," *New York Times*, June 29, 1977; Kristin McMurran, "Shirley Temple Black Taps Out a Telling Memoir of Child Stardom," *People*, November 28, 1988; "The Prima Donna," *Time*, October 29, 1956; Lily Rothman, "Ambassador Shirley Temple," *Time*, February 13, 2014; Edwin Schallert, "Child Carries Starring Role," *Los Angeles Times*, May 26, 1934; Schallert, "Child Shines at Loew's," *Los Angeles Times*, June 29, 1934; "Shirley Temple Grows Up," *Life*, March 30, 1942; "Shirley Opens Storybook," *Life*, February 3, 1958; "Shirley Temple and the Pearls," *New York Times*, June 30, 1934; "Shirley Temple Pursued: 8,000 Children Nearly Mob Actress on Arrival at Honolulu," *New York Times*, July 30, 1935; Margaret Talbot, "Shirley Temple: Our Little Girl," *New Yorker*, February 13, 2014; Michael Taylor, "Charles Black, the Man Behind Shirley Temple," *San Francisco Chronicle*, August 9, 2005; Claire Trask, "The Screen in Berlin," *New York Times*, April 28, 1935; Ida Zeitlin, "Gable Is Afraid of Shirley Temple!" *Modern Screen*, August 1937.

Additional Resources
"Shirley Temple: America's Little Darling," The Hollywood Collection, directed by Gene Feldman, Janson Media, June 30, 2016; Shirley Temple Black, interview by Dick Foley and Dana Middleton, *Northwest Afternoon*, KOMO TV, Seattle, November 14, 1988; Shirley Temple Black, interview by Jane Pauley, *The Today Show*, NBC, October 1988; Shirley Temple Black, interview by Larry King, *Larry King Live*, CNN, October 25, 1988; Shirley Temple Black, interview by Regis

Philbin and Kathie Lee, *Live With Regis & Kathie Lee*, ABC, October 25, 1988; Shirley Temple, Eleanor Roosevelt, "My Day," March 19, 1938 and Roosevelt, "My Day," July 11, 1938, accessed on *www2.gwu.edu/~erpapers/myday/browse byyear.cfm.*

CHAPTER 3: Yo-Yo Ma
Books

Pablo Casals, *Joys and Sorrows,* as told to Albert E. Kahn (New York: Simon & Schuster, 1970); Henry Louis Gates, Jr., *Faces of America: How 12 Extraordinary People Discovered Their Pasts* (New York and London: New York University Press, 2010)*; Kay Redfield Jamison, *Exuberance: The Passion for Life* (New York: Vintage Books, 2005); Claude Kenneson, *Musical Prodigies: Perilous Journeys, Remarkable Lives* (Portland, OR: Amadeus Press, 1998); Marina Ma, *My Son, Yo-Yo,* as told to John A. Rallo (Hong Kong: The Chinese University of Hong Kong Press, 1996); Gary E. McPherson, ed., *Musical Prodigies: Interpretations From Psychology, Education, Musicology, and Ethnomusicology* (Oxford, UK: Oxford University Press, 2016); Eric Siblin, *The Cello Suites: J.S. Bach, Pablo Casals and the Search for a Baroque Masterpiece* (New York: Grove Press, 2009); Isaac Stern, *My First 79 Years,* written with Chaim Potok (New York: Alfred A. Knopf, 1999); *The Letters of Mozart and His Family,* edited by Emily Anderson (New York: Macmillan Press, 1989)*; Megan Westberg, ed., *Yo-Yo Ma: Portrait of a Cellist, Mentor & Musical Explorer* (San Anselmo, CA: Stringletter, 2012).

Newspapers, Magazines, and Scientific Journals

Karen Chan Barrett et al., "Classical Creativity: A Functional Magnetic Resonance Imaging (fMRI) Investigation of Pianist and Improviser Gabriela Montero," *NeuroImage* 209 (April 1, 2020); David Blum, "A Process Larger Than Oneself," *New Yorker,* May 1, 1989; Benjamin C. Burns, "Yo-Yo Ma Goes Beyond the Music," *Harvard Crimson,* February 12, 2009; Gilles Comeau et al., "Can You Tell a Prodigy From a Professional Musician?" *Music Perception* 35, No. 2 (December 2017), 200-209; Richard Dyer, "Yo-Yo Ma: A Debut at Last in His

Spark

Own Words," *Boston Globe,* February 6, 1983; Helen Epstein, "Yo-Yo Ma, The Serene and Vibrant Life of a Virtuoso," *Esquire,* March 1982; Arthur Gelb, "Spectacle on Closed-Circuit TV to Herald Cultural Center Drive," *New York Times,* November 19, 1962; Richard S. Ginell, "Yo-Yo Ma, Edgar Meyer, Mark O'Connor," *Variety,* November 7, 1996; David Z. Hambrick, "What Makes a Prodigy?" *Scientific American,* September 22, 2015; Joseph Horowitz, "Exuberance Plus Serenity Equals Yo-Yo Ma," *New York Times,* April 15, 1979; Allen Hughes, "Yo-Yo Ma, Cellist, Now 16, Performs an Enviable Recital," *New York Times,* May 8, 1971; Jamie James, "Yo-Yo Ma May Be a National Institution, but He Continues to Reinvent Himself," *New York Times,* December 31, 1995; Scott Barry Kaufman, "Opening Up Openness to Experience," *Journal of Creative Behavior* 47, No. 4 (2013), 233-255; Frances Lewine, "Closed TV Program Opens Culture Crusade," Associated Press in *The Journal Times,* November 30, 1962; Daniel J. Levitin and Scott T. Grafton, "Measuring the Representational Space of Music with fMRI: A Case Study With Sting," *Neurocase* 22, No. 6 (2016), 548-557; James McCarthy, "Yo-Yo Ma, Interview by Philip Kennicott," *Gramophone,* May 3, 2013; Gary E. McPherson, "The Role of Parents in Children's Musical Development," *Psychology of Music* 37, No. 1 (January 2009), 91-110; James R. Oestreich, "At Lunch With: Yo-Yo Ma; 1 Part Earnestness, 1 Blast of Laughter," *New York Times,* September 21, 1994; Gottfried Schlaug, "Musicians and Music Making as a Model for the Study of Brain Plasticity," *Progress in Brain Research* 217 (2015), 37-55; Martin Steinberg, "Magic Cello Ride," *Strings* 247 (November 2015); Bernard Taper, "A Cellist in Exile," *New Yorker,* February 24, 1962; Janet Tassel, "Yo-Yo Ma's Journeys," *Harvard Magazine,* March-April 2000; Richard Thorne, "The Magic of Yo-Yo Ma," *Saturday Review,* July 1981; Darold Treffert, "The Savant Syndrome: An Extraordinary Condition. A Synopsis: Past, Present, Future," *Philosophical Transactions of the Royal Society B: Biological Sciences* 364 (1522) (May 27, 2009); Larry R. Vandervert, "The Appearance of the Child Prodigy 10,000 Years Ago: An Evolutionary and Developmental Explanation," *Journal of Mind and Behavior* 30, Nos. 1 and 2 (Winter and Spring 2009), 15-32; Vandervert, "How Music

Training Enhances Working Memory: A Cerebrocerebellar Blending Mechanism That Can Lead to Scientific Discovery and Therapeutic Efficacy in Neurological Disorders," *Cerebellum & Ataxias* 2, No. 11 (2015).

Additional Resources

Yo-Yo Ma, interview by Soledad O'Brien, *Fireside Chat With Trailblazer Yo-Yo Ma*, Salesforce, June 17, 2019; Yo-Yo Ma, interview by David Rubenstein, *The David Rubenstein Show: Peer to Peer Conversations*, PBS, September 27, 2019; Yo-Yo Ma, *Classic Yo-Yo Ma*, directed by Jennifer Warner, Amazon Prime, 2001.

Chapter 4: Bill Gates
Books

Paul Allen, *Idea Man: A Memoir by the Cofounder of Microsoft* (New York: Portfolio/Penguin, 2011)*; Bill Gates, with Nathan Myhrvold and Peter Rinearson, *The Road Ahead* (New York: Viking, 1995)*; Bill Gates, Sr., with Mary Ann Mackin, *Showing Up for Life: Thoughts on the Gifts of a Lifetime* (New York: Broadway Books, 2009); Melinda Gates, *The Moment of Lift: How Empowering Women Changes the World* (New York: Flatiron Books, 2019); Malcolm Gladwell, *Outliers: The Story of Success* (New York: Little, Brown and Co., 2008); Walter Isaacson, *The Innovators: How a Group of Hackers, Geniuses, and Geeks Created the Digital Revolution* (New York: Simon & Schuster, 2014); Stephen Manes and Paul Andrews, *Gates: How Microsoft's Mogul Reinvented an Industry—and Made Himself the Richest Man in America* (New York: Touchstone, 1994); Keith Sawyer, *Group Genius: The Creative Power of Collaboration* (New York: Basic Books, 2017); James Wallace and Jim Erickson, *Hard Drive: Bill Gates and the Making of the Microsoft Empire* (New York: HarperCollins, 1993).

Newspapers, Magazines, and Scientific Journals

Paul Andrews, "Mary Gates: She's Much More Than Just the Mother of Billionaire Bill," *Seattle Times,* January 9, 1994; "From 7 Science Laureates: The Teamwork Behind a Nobel," Letter to the Editor, *New York Times,* October 4, 2017;

Bill Gates, "Remarks of Bill Gates, Harvard Commencement 2007," *Harvard Gazette,* June 7, 2007; Melinda Gates, "The Vacation That Changed Everything," *AARP,* November 28, 2017; Julianne Holt-Lunstad et al., "Social Relationships and Mortality Risk: A Meta-Analytic Review," *PLoS Medicine* 7 (7) (July 2010); Nicholas D. Kristof, "For Third World, Water Is Still a Deadly Drink," *New York Times,* January 9, 1997; Norman Maier, "Reasoning in Humans," *Journal of Comparative Psychology* 12, No. 2 (1931), 181-194; Betsy McKay, "Bill Gates Has Regrets," *Wall Street Journal,* May 11, 2020; Olga Stavrova, "Having a Happy Spouse Is Associated With Lower Risk of Mortality," *Psychological Science* 30, No. 5 (May 2019), 798-803; "Tribute From Miss Anthony," *New York Times,* October 27, 1902.

Additional Resources
Warren Buffett and Bill Gates, "Warren Buffett and Bill Gates: Keeping America Great," interview by Betsy Quick, CNBC, November 12, 2009; *forbes.com/profile/bill-gates/#6c26e729689f;* Bill Gates, interview by Charlie Rose, *The Charlie Rose Show,* PBS, February 22, 2016; Bill and Melinda Gates, "Pop Quiz with Bill and Melinda Gates," *The Late Show with Stephen Colbert,* CBS, February 19, 2019; Bill and Melinda Gates, "Why Giving Away Our Wealth Has Been the Most Satisfying Thing We've Done," interview by Chris Anderson, TED 2014, March 2014; "Inside Bill's Brain: Decoding Bill Gates," directed by Davis Guggenheim. Netflix, 2019; *gatesfoundation.org; givingpledge.org; lakesideschool.org; reddit.com; ted.com.*

CHAPTER 5: Isaac Newton
Books
Peter Ackroyd, *Isaac Newton* (London: Vintage Books, 2007); Patricia Fara, *Newton: The Making of Genius* (London: Picador, 2003); Fara, *Isaac Newton at Woolsthorpe Manor* (Swindon, UK: Park Lane Press, 2018); James Gleick, *Isaac Newton* (New York: Vintage Books, 2003); Rob Iliffe, *Newton: A Very Short Introduction* (Oxford, UK: Oxford University Press, 2007); Oliver Sacks, *On the*

Sources and Notes

Move: A Life (New York and Toronto: Alfred A. Knopf, 2015)*; Paul Strathern, *Mendeleyev's Dream: The Quest for the Elements* (New York: Pegasus Books, 2019); Henry David Thoreau, *The Portable Thoreau*, edited by Jeffrey S. Cramer (New York: Penguin Books, 2012); Richard S. Westfall, *The Life of Isaac Newton* (Cambridge, UK: Cambridge University Press, 1993).

Newspapers, Magazines, and Scientific Journals

Paul R. Heyl, "Why Newton Is Remembered," *Science News-Letter* XI (No. 310), March 19, 1927; L.W. Johnson and M.L. Wolbarsht, "Mercury Poisoning a Probable Cause of Isaac Newton's Physical and Mental Ills," *Notes and Records of the Royal Society of London* 34, No.1 (July 1979), 1-9; Milo Keynes, "The Personality of Isaac Newton," *Notes and Records of the Royal Society of London* 49, No. 1 (January 1995), 1-56; John Kounios and Mark Beeman, "The Cognitive Neuroscience of Insight," *Annual Review of Psychology* 65 (2014), 71-93; Kounios and Beeman, "The Aha! Moment," *Current Directions in Psychological Science* 18, No. 4 (August 1, 2009), 210-216; Daniel Kuehn, "Keynes, Newton and the Royal Society," *Notes and Records of the Royal Society* 67, No. 1 (March 20, 2013), 25-36; Harvey C. Lehman, "The Chemist's Most Creative Years," *Science* 127, No. 3308 (May 23, 1958), 1213-1222; James R. Newman, "A Fine Collection of Essays About Isaac Newton, Both the Man and His Accomplishments, on the Occasion of His Tercentenary," *Scientific American* 179, No. 1 (July 1948), 56-59; Robert Sinatra, Albert-László Barabási et al., "Quantifying the Evolution of Individual Scientific Impact," *Science* 354, No. 6312 (November 4, 2016); P. E. Spargo and C. A. Pounds, "Newton's Derangement of the Intellect: New Light on an Old Problem," *Notes and Records of the Royal Society* 34, No.1 (July 1979), 11-32; Anthony Storr, "Isaac Newton," *British Medical Journal* 291, No. 6511 (December 21-28, 1985), 1779-1784.

Additional Resources

Albert-László Barabási, "The Real Relationship Between Your Age and Your Chance of Success," March 30, 2019, Washington, DC, TEDxMidAtlantic,

16:08, *ted.com/talks/albert_laszlo_barabasi_the_real_relationship_between_your_age_and_your_chance_of_success;* "Beautiful Minds: The Enigma of Genius," June 4, 2011, New York, World Science Festival, 01:30:32, *worldsciencefestival.com/videos/beautiful-minds-the-enigma-of-genius;* Elizabeth Gilbert, "Your Elusive Creative Genius," February 5, 2009, Long Beach, CA, TED 2009 Podcast, 19:16, *ted.com/talks/elizabeth_gilbert_your_elusive_creative_genius; cudl.lib.cam.ac.uk/collections/newton/1; nationaltrust.org.uk/woolsthorpe-manor; newtonproject.ox.ac.uk; royalmint.com; worldsciencefestival.com.*

CHAPTER 6: **Sara Blakely**
Newspapers, Magazines, and Scientific Journals
Pierre Azoulay et al., "Age and High-Growth Entrepreneurship," *American Economic Review: Insights* 2, No. 1 (2020), 65-82; Jessica Bennett, "On Campus, Failure Is on the Syllabus," *New York Times,* June 24, 2017; Emily Sakai, "New Kellogg Research Shows Early Career Failures Lead to Future Success," *Daily Northwestern,* October 14, 2019; Xiaodong Lin-Siegler et al., "Even Einstein Struggled," *Journal of Educational Psychology* 108, No. 3 (2016), 314-328; Yang Wang, Benjamin F. Jones, and Dashun Wang, "Early-Career Setback and Future Career Impact," *Nature Communications* 10, No. 4331 (2019).

Additional Resources
"How Sara Blakely Created Spanx with $5,000," *The Oprah Winfrey Show,* OWN, February 1, 2007, *oprah.com/own-oprahshow/how-sara-blakely-created-spanx-with-5000;* "Spanx: Sara Blakely," interview by Guy Raz, *How I Built This With Guy Raz* Podcast, NPR, September 12, 2016, *npr.org/2017/08/15/534771839/spanx-sara-blakely.*

CHAPTER 7: **Julia Child**
Books
Nancy Verde Barr, *Backstage With Julia: My Years With Julia Child* (Hoboken, NJ: John Wiley & Sons, 2007); *Julia Child: The Last Interview and Other*

Sources and Notes

Conversations (Brooklyn, NY: Melville House Publishing, 2019); Julia Child and Alex Prud'homme, *My Life in France* (New York: Anchor Books, 2007); Julia Child, Louisette Bertholle, and Simone Beck, *Mastering the Art of French Cooking* (New York: Knopf, 1989); Noel Riley Fitch, *Appetite for Life: The Biography of Julia Child* (New York: Doubleday, 1997)*; Ernest Hemingway, *A Moveable Feast* (New York: Charles Scribner's Sons, 1964)*; Judith Jones, *The Tenth Muse: My Life in Food* (New York: Alfred A. Knopf, 2007); Alex Prud'homme, *The French Chef in America: Julia Child's Second Act* (New York: Anchor Books, 2017); Prud'homme and Katie Pratt, *France Is a Feast: The Photographic Journey of Paul and Julia Child* (New York: Thames & Hudson, 2017); Joan Reardon, ed., *As Always, Julia: The Letters of Julia Child & Avis Devoto* (New York: Mariner Books, 2012); Laura Shapiro, *Julia Child* (New York: Viking, 2007); Bob Spitz, *Dearie: The Remarkable Life of Julia Child* (New York: Vintage Books, 2013).

Newspapers, Magazines, and Scientific Journals

Pamela Druckerman, "How the Midlife Crisis Came to Be," *Atlantic,* May 29, 2018; "Everyone's in the Kitchen," *Time,* November 25, 1966; Stephanie Hersh, "A Full Measure of Humor," *Gastronomica* 5, No. 3 (Summer 2005), 15-17; Elliott Jacques, "Death and the Mid-Life Crisis," *International Journal of Psychoanalysis* 46, No. 4 (October 1965), 502-514; Paula Johnson, "Remembering Julia Child," *Smithsonian,* August 13, 2019; Paul A. O'Keefe, Carol S. Dweck, and Gregory M. Walton, "Implicit Theories of Interest: Finding Your Passion or Developing It?" *Psychological Science* 29, No. 10 (October 1, 2018), 1653-1664; Jacques Pépin, "My Friend Julia Child," *Gastronomica* 5, No. 3 (Summer 2005), 9-14; Gail Cameron Wescott, "Mastering the Art of Living," *Smith Alumnae Quarterly* (Winter 2002-2003), 39-43.

Additional Resources

americanhistory.si.edu; juliachildfoundation.org; "Alice Waters: I Was Educated in French Restaurants," interview by Clément Thiery, France-Amérique, October 26, 2017, *france-amerique.com; wgbh.org/foundation/julia-child.*

CHAPTER 8: Maya Angelou
Books

Maya Angelou, *Letter to My Daughter* (New York: Random House, 2008); Angelou, *Mom & Me & Mom* (New York: Random House, 2013)*; Angelou, *The Collected Autobiographies of Maya Angelou* (New York: Random House, 2004); Frederick Douglass, *Narrative of the Life of Frederick Douglass, An American Slave* (Boston: Anti-Slavery Office, 1845)*; Marcia Ann Gillespie, Rosa Johnson Butler, and Richard A. Long, *Maya Angelou: A Glorious Celebration* (New York: Doubleday, 2008); Elizabeth Keckley, *Thirty Years a Slave, and Four Years in the White House* (New York: G.W. Carleton & Co., 1868)**; *Twenty-four Conversations with Borges: Including a Selection of Poems,* Interviews by Roberto Alifano, 1981-1983 (Housatonic, MA: Lascaux Publishers, 1984).

Newspapers, Magazines, and Scientific Journals

Hilton Als, "Songbird," *New Yorker,* August 5, 2002; J. Marvin Eisenstadt, "Parental Loss and Genius," *American Psychologist* (March 1978), 211-223; Ward Just, "'The Caged Bird' Sings of a Black Childhood," *Washington Post,* April 3, 1970; Amy Longsdorf, "'Delta' Is the Dawn of Another Career for Author Maya Angelou," *Morning Call,* December 27, 1998; Mary Jane Lupton, "'When Great Trees Fall': The Poetry of Maya Angelou," *CLA Journal* 58, No. 1/2 (September/December 2014), 77-90; "Maya Angelou," *Wake Forest Magazine,* Fall 2014; Audrey T. McCluskey, "Maya Angelou: Telling the Truth, Eloquently," *Black Camera* 16, No. 2 (Fall/Winter 2001), 3-4, 11; Carol E. Neubauer, "An Interview With Maya Angelou," *The Massachusetts Review* 28, No. 2 (Summer 1987), 286-292; "On Her Third Attempt, Maya Angelou Finds the Ideal Husband, Germaine Greer's Ex," *People,* August 25, 1975; George Plimpton, "Maya Angelou: The Art of Fiction," *The Paris Review* 116 (Fall 1990); Hayley Robinson et al., "The Effects of Expressive Writing Before or After Punch Biopsy on Wound Healing," *Brain, Behavior, and Immunity* 61 (March 2017), 217-227; David Ulin, "When Maya Angelou Was Interviewed by Studs Terkel, *Los Angeles Times,* May 28, 2014.

Sources and Notes

Additional Resources

"Maya Angelou: And Still I Rise," directed by Bob Hercules and Rita Coburn Whack, *American Masters,* PBS, February 21, 2017; Maya Angelou, interview by Oprah Winfrey, "Conversations With Oprah: Maya Angelou," *The Oprah Winfrey Show,* July 13, 1993; Maya Angelou, interviewed by Bill Moyers, "Going Home with Maya Angelou," *Moyers & Company,* PBS, August 7, 2014; Maya Angelou, interviewed by Brian Lehrer, "Maya Angelou: Her Two Moms," *The Brian Lehrer Show,* WNYC, April 2, 2013; "Maya Angelou: Learning to Love My Mother," BBC News, April 3, 2013; Maya Angelou, interviewed by Oprah Winfrey, "Soul to Soul with Dr. Maya Angelou," *SuperSoul Sunday,* OWN, May 28, 2014.

CHAPTER 9: Alexander Fleming
Books

Kevin Brown, *Penicillin Man: Alexander Fleming and the Antibiotic Revolution* (Cheltenham, UK: The History Press, 2005); Gwyn Macfarlane, *Alexander Fleming: The Man and the Myth* (Cambridge, MA: Harvard University Press, 1984); André Maurois, *The Life of Sir Alexander Fleming: Discoverer of Penicillin* (New York: E.P Dutton and Co., 1959); Royston M. Roberts, *Serendipity: Accidental Discoveries in Science* (New York: John Wiley & Sons, 1989).

Newspapers, Magazines, and Scientific Journals

V. D. Allison, "Personal Recollections of Sir Almroth Wright and Sir Alexander Fleming," *Ulster Medical Journal* 43, No. 2 (1974), 89-98; James H. Austin, "The Roots of Serendipity," *Saturday Review,* November 2, 1974, 60-64; Albert-László Barabási, "Don't Give Up: Older People Can Have Creative Breakthroughs," *Washington Post,* December 10, 2016; L. Colebrook, "Alexander Fleming: 1881-1955," *Biographical Memoirs of Fellows of the Royal Society* 2 (November 1956), 117-126; "Obituary: Sir Alexander Fleming, M.D., D.Sc., F.R.C.P., F.R.C.S., F.R.S," *British Medical Journal* 1, No. 4915 (March 19, 1955), 732-735; Benjamin F. Jones and Bruce A. Weinberg, "Age Dynamics in

Scientific Creativity," *Proceedings of the National Academy of Sciences* 108, No. 47 (November 22, 2011), 18910-18914; Claudia Kalb, "Remembering Cancer Researcher Judah Folkman," *Newsweek,* January 15, 2008; Wolfgang Saxon, "Anne Miller, 90, First Patient Who Was Saved by Penicillin," *New York Times,* June 9, 1999; Arthur Silverstein, "The Plato of Praed Street: The Life and Times of Almroth Wright," *Bulletin of the History of Medicine* 76, No. 1 (Spring 2002), 158-160; S. J. Woolf, "Sir Alexander Fleming—Man of Science and of Penicillin," *American Scientist* 33, No. 4 (October 1945), 242-245, 252.

Additional Resources

"St. Mary's Hospital Honours Sir Alexander Fleming, 1954," *britishpathe.com/ video/st-marys-hospital-honours-sir-alexander-fleming;* "The Big Bang's Echo," interview by Ralph Schoenstein, *All Things Considered,* NPR, May 17, 2005, *https://www.npr.org/templates/story/story.php?storyId=4655517;* "The Large Horn Antenna and the Discovery of Cosmic Microwave Background Radiation," American Physical Society, *aps.org/programs/outreach/history/historicsites/penzias wilson.cfm;* Alexander Fleming Laboratory Museum, *imperial.nhs.uk/about-us/ who-we-are/fleming-museum.*

CHAPTER 10: Eleanor Roosevelt
Books

Bernard Asbell, ed., *Mother and Daughter: The Letters of Eleanor and Anna Roosevelt* (New York; Fromm International Publishing, 1988); Peter Collier with David Horowitz, *The Roosevelts: An American Saga* (New York: Simon & Schuster, 1994); Blanche Wiesen Cook, *Eleanor Roosevelt, Volume 1: The Early Years, 1884-1933* (New York: Penguin Books, 1993); Cook, *Eleanor Roosevelt, Volume 2: The Defining Years, 1933-1938* (New York: Penguin Books, 1999); Cook, *Eleanor Roosevelt, Volume 3: The War Years and After, 1939-1962* (New York: Penguin Books, 2017); Yael Danieli, Elsa Stamatopoulou, and Clarence J. Dias, eds., *The Universal Declaration of Human Rights: Fifty Years and Beyond,* (Amityville, NY: Baywood Publishing Co., 1999)*; Lorena A. Hickok, *Eleanor Roosevelt: Reluctant*

Sources and Notes

First Lady (New York: Dodd, Mead & Co., 1980); Joseph P. Lash, *Eleanor: The Years Alone* (New York: Smithmark Publishers, 1972); Lash, *Eleanor and Franklin: The Story of their Relationship Based on Eleanor Roosevelt's Private Papers* (New York: W.W. Norton & Co., 1971); David B. Roosevelt, with Manuela Dunn-Mascetti, *Grandmère: A Personal History of Eleanor Roosevelt* (New York: Warner Books, 2002); Eleanor Roosevelt, *The Autobiography of Eleanor Roosevelt* (New York: HarperCollins, 2014); Roosevelt, *You Learn by Living: Eleven Keys for a More Fulfilling Life* (New York: HarperCollins, 2016); Roosevelt, *Eleanor Roosevelt in Her Words: On Women, Politics, Leadership, and Lessons From Life*, edited by Nancy Woloch (New York: Black Dog & Leventhal, 2017); Roosevelt, *My Day: The Best of Eleanor Roosevelt's Acclaimed Newspaper Columns, 1936-1962* (New York: MJF Books, 2001); Elliott Roosevelt, *Hunting Big Game in the Eighties: The Letters of Elliott Roosevelt, Sportsman*, edited by Anna Eleanor Roosevelt (New York: Charles Scribner's Sons, 1933)*; James Roosevelt, *My Parents: A Differing View* (New York: Playboy Press, 1976); *The Wisdom of Eleanor Roosevelt: McCall's Tribute to an Illustrious Woman* (New York: McCall Corporation, January 1962).

Newspapers, Magazines, and Scientific Journals
Rudy A. Black, "Mrs. Roosevelt Takes Night Ride in Plane Piloted by Amelia Earhart," United Press International in *Salt Lake Telegram*, April 21, 1933; Sarah Booth Conroy, "In Eleanor Roosevelt's Orbit," *Washington Post*, March 11, 1996; "Elliott Roosevelt Insane," *New York Times*, August 18, 1891; "First Lady Tours Coal Mine in Ohio," *New York Times*, May 22, 1935; "A Great Society Event: The Ball Given by Mrs. William Astor," *New York Times*, January 22, 1884; Arthur Krock, "President Roosevelt Is Dead; Truman to Continue Policies," *New York Times*, April 12, 1945; Barron H. Lerner, "Final Diagnosis," *Washington Post*, February 8, 2000; "The Obituary Record: Elliott Roosevelt," *New York Times*, August 16, 1894; "President Roosevelt Gives the Bride Away: His Niece Weds His Cousin, Franklin Delano Roosevelt," *New York Times*, March 18, 1905; Eleanor Roosevelt, "In Defense of Curiosity," *Saturday Evening Post*, August 24,

1935; Scott J. Russo et al., "Neurobiology of Resilience," *Nature Neuroscience* 15, No. 11 (November 2012), 1475-1484; Martin Teicher et al., "The Effects of Childhood Maltreatment on Brain Structure, Function and Connectivity," *Nature Reviews: Neuroscience* 17 (October 2016), 652-666; Emmy E. Werner, "Children of the Garden Island," *Scientific American* 260, No. 4 (April 1989), 106-111; Werner, "Resilience and Recovery: Findings From the Kauai Longitudinal Study," *Focal Point* 19, No. 1 (Summer 2005), 11-14.

Additional Resources

archives.gov/exhibits/american_originals/eleanor; fdrlibrary.org; nps.gov/elro; theshipslist.com/ships/Wrecks/brit&celtic.shtml; vault.fbi.gov/Eleanor%20Roosevelt.

CHAPTER 11: Peter Mark Roget
Books

George W. Bush, *Decision Points* (New York; Crown, 2010)*; Humphry Davy, *Researches, Chemical and Philosophical, Chiefly Concerning Nitrous Oxide, or Dephlogisticated Nitrous Air, and Its Respiration* (London: Biggs and Cottle, 1800)*; Harriett Doerr, *Stones for Ibarra* (New York: Penguin Books, 1984); D. L. Emblen, *Peter Mark Roget: The Word and the Man* (New York: Thomas Y. Crowell Co., 1970); Kay Redfield Jamison, *Robert Lowell, Setting the River on Fire: A Study of Genius, Mania, and Character* (New York: Vintage Books, 2018); Joshua Kendall, *The Man Who Made Lists: Love, Death, Madness, and the Creation of* Roget's Thesaurus (New York: Berkley Books, 2009); Herbert Philips, *Continental Travel in 1802–3: The Story of an Escape* (Manchester, UK: Sherratt & Hughes, 1904)**; Nick Rennison, *Peter Mark Roget: The Man Who Became a Book* (Pocket Essentials series) (Harpenden, UK: Oldcastle Books, 2007); Peter Mark Roget, *Roget's Thesaurus of English Words and Phrases* (New York: Grosset & Dunlap, 1941); Peter Mark Roget, *Animal and Vegetable Physiology, Considered With Reference to Natural Theology* (Philadelphia: Carey, Lea & Blanchard, 1836)**; Samuel Romilly, *Memoirs of the Life of Sir Samuel Romilly* (London: John Murray, 1840)**; Virginia Woolf, *The Diary of Virginia Woolf: Volume Two,*

Sources and Notes

1920-1924, edited by Anne Olivier Bell, assisted by Andrew McNeillie (New York: Harcourt Brace Jovanovich, 1978).*

Newspapers, Magazines, and Scientific Journals

Alison Beard, "Life's Work: An Interview With Maya Angelou," *Harvard Business Review*, May 2013; Anatole Broyard, "Books of the Times," *New York Times*, December 7, 1983; George H. Douglas, "What's Happened to the Thesaurus?" *Reference Quarterly* 16, No. 2 (Winter 1976), 149-155; Letters to the Editor, *Atlantic Monthly*, September 2001; William S. Odlin, "Roget Becomes Saint of Crosswordia," *New York Times*, February 8, 1925; Peter Mark Roget, "Age," *The Cyclopaedia of Practical Medicine* 1 (1845); Samuel Romilly death, *The Times*, November 3, 1818; Alex Ross, "György Kurtág, With His Opera of 'Endgame,' Proves to Be Beckett's Equal," *New Yorker*, December 24 & 31, 2018; "Royal Institution: Fullerian Professorship," *London Medical and Surgical Journal* 1, 1837; Oliver Sacks, "Henry Cavendish: An Early Case of Asperger's Syndrome?" *Neurology* 57 (October 2001), 1347; Wallace Stegner, "Women We Love: Harriet Doerr," *Esquire*, December 1988; William E. Swinton, "The Remarkable Accomplishments of Dr. Peter Roget," *CMA Journal* 123 (November 8, 1980), 916-918, 921; *Westminster Review* 59 (January-April 1853)**; Simon Winchester, "Word Imperfect," *Atlantic*, May 2001; Ben Zimmer, "Treasure Trove," *New York Times Magazine*, January 7, 2011; Ben Zimmer, "Word for Word," *Lapham's Quarterly* 5, No. 2 (Spring 2012).

CHAPTER 12: Grandma Moses

Books

David W. Galenson, *Old Masters and Young Geniuses: The Two Life Cycles of Artistic Creativity* (Princeton, NJ: Princeton University Press, 2007); Jane Kallir, *Grandma Moses in the 21st Century* (Alexandria, VA: Art Services International, 2001); Otto Kallir, *Grandma Moses* (New York: Harry N. Abrams, 1973); Otto Kallir, ed., *Grandma Moses: My Life's History* (New York: Harper & Brothers, 1952); Oliver Sacks, *Musicophilia: Tales of Music and the Brain* (New York: Alfred A. Knopf, 2007).*

Spark

Newspapers, Magazines, and Scientific Journals
Aliya Alimujiang et al., "Association Between Life Purpose and Mortality Among US Adults Older Than 50 Years," *JAMA Network Open* 2, No. 5 (2019); Maura Boldrini et al., "Human Hippocampal Neurogenesis Persists Throughout Aging," *Cell Stem Cell* 22 (April 5, 2018), 589-599; Thomas Buckley, "Woman Botanist Still Busy at 79," *New York Times,* May 29, 1963; "Gift Painting of His Farm (Grandma Moses' Version) Delights President," *New York Times,* January 19, 1956; Hilda Goodwin, "Grandma Moses, Eagle Bridge Artist, Dies at 101," *Times Record,* December 14, 1961; Goodwin, "Grandma Moses, Impatient With Orders to Rest, Hides Doctor's Stethoscope," *Times Record,* August 31, 1961; Pagan Kennedy, "To Be a Genius, Think Like a 94-Year-Old," *New York Times,* April 7, 2017; "Lady Botanist," *New Yorker,* July 27, 1963; "100 Candles for a Gay Lady: Grandma Moses," *Life,* September 19, 1960; "100th Birthday Tributes Come From All Over the World," *Times Record,* September 3, 1960; "President Greets Grandma Moses," *Times Record,* September 3, 1960; Eleanor Roosevelt, "My Day," *Capital Times,* May 17, 1949; Oliver Sacks, "The Joy of Old Age. (No Kidding.)," *New York Times,* July 6, 2013; Peter Schjeldahl, "The Original: Grandma Moses Looks Better Than Ever," *New Yorker,* May 28, 2001; Katharine Q. Seelye, "Barbara Hillary, 88, Trailblazer on Top (and Bottom) of the World, Dies," *New York Times,* November 26, 2019; Helen Starkweather, "Interview: David Galenson," *Smithsonian,* November 2006; S. J. Woolf, "Grandma Moses, Who Began to Paint at 78," *New York Times,* December 2, 1945.

Additional Resources
Anna Mary Robertson Moses, interviewed by Edward R. Murrow, "See It Now: Two American Originals," CBS, December 12, 1955; "Game App Developer in Her 80s Opens ICT World for Fellow Seniors, *japan.go.jp/tomodachi/2018/spring -summer2018/game_app_developer.html.; www.nga.gov.*

EPILOGUE: Leonardo da Vinci
This chapter includes material from a cover story I wrote for *National Geographic:*

Sources and Notes

"Leonardo da Vinci's Brilliance Endures 500 Years after his Death," published in May 2019.

Books

Carmen C. Bambach, ed., *Leonardo da Vinci: Master Draftsman* (New York: Metropolitan Museum of Art, 2003); Bambach, *Leonardo Rediscovered* (New Haven, CT: Yale University Press, 2019); Martin Clayton and Ron Philo, *Leonardo da Vinci: Anatomist* (London: Royal Collection Trust, 2014); Clayton, *Leonardo da Vinci: A Life in Drawing* (London: Royal Collection Trust, 2019); Walter Isaacson, *Leonardo da Vinci* (New York: Simon & Schuster, 2017); Martin Kemp, *Leonardo* (New York: Oxford University Press, 2011); Kemp, *Living With Leonardo: Fifty Years of Sanity and Insanity in the Art World and Beyond* (London: Thames & Hudson, 2018); Stefan Klein, *Leonardo's Legacy: How Da Vinci Reimagined the World,* translated by Shelley Frisch (Cambridge, MA: Da Capo Press, 2010); *Leonardo da Vinci Notebooks,* selected by Irma A. Richter, edited by Thereza Wells (Oxford, UK: Oxford University Press, 2008); Sherwin B. Nuland, *Leonardo da Vinci* (New York: Penguin, 2000); H. Anna Suh, ed., *Leonardo's Notebooks: Writing and Art of the Great Master* (New York: Black Dog & Leventhal, 2005); Alessandro Vezzosi, *Leonardo da Vinci: Renaissance Man* (London: Thames & Hudson, 2016); Francis Wells, *The Heart of Leonardo* (London: Springer-Verlag, 2013); Emanuel Winternitz, *Leonardo da Vinci as a Musician* (New Haven, CT: Yale University Press, 1982).

*Books accessed on The Internet Archive *(archive.org)*
**Books accessed on HathiTrust Digital Library *(hathitrust.org)*

Grateful acknowledgment is made for permission to reprint portions of previously published material:
Excerpt(s) from *Gather Together in My Name* by Maya Angelou, copyright © 1974 by Maya Angelou. Used by permission of Penguin Random House LLC and Little, Brown Book Group. All rights reserved.

Spark

Acknowledgments

O ne summer, many years ago, I spent several glorious weeks as a college English major studying literature at Trinity College, Oxford. I have wonderful memories of reading prose and poetry, walking the centuries-old streets, and punting on the River Cherwell. But my most vivid recollection is this: seeing my Oxford don, Christopher Ricks, listening to Bob Dylan before class with his eyes shut.

It seemed an odd confluence to me at the time—the eminent T. S. Eliot scholar lost in Dylan's gravelly voice. But I soon learned that Ricks was a Dylan devotee who had discovered the richness of the musician's songwriting one night at a dinner party. When the host turned out the lights and played "Desolation Row," Ricks listened to Dylan—really *listened*—for the first time, he said. Later, Ricks called Dylan "the greatest living user of the English language."

This image of Ricks and Dylan sprang to mind when Keith Sawyer, whose research I discuss in Chapter 4, mentioned Eliot and his poem *The Waste Land* in the course of our conversation about collaboration. I decided to look up the lyrics to "Desolation Row" and was wowed by the number of references to characters—biblical, fictional, and real—that Dylan threads through his song: Cinderella, Bette Davis, Romeo, Cain and Abel, Ophelia, Noah, Casanova, Einstein, and, yes, Eliot and his editor, Ezra Pound. Suddenly, "Desolation Row" struck me as a wonderfully apt metaphor for the process required of any piece of writing. Without these personas—collaborators in their own unique way—Dylan's music would have been a wasteland of its own, devoid of the souls that make his lyrics sing.

Collaboration fuels creativity in all of us. I am enormously grateful to the cadre of people who made my writing of this book possible. First and foremost, I thank Hilary Black, my editor at National Geographic Books, who guided me through this project with abundant wisdom and support. I lucked into a most rewarding partnership with Hilary on my first book and am beyond grateful that we had the opportunity to work together again. Hilary's enthusiasm for the subject matter, her editorial sagacity, and her kindness and championing of my work mean everything.

I am privileged to be a National Geographic author and owe enormous gratitude to the fantastic editorial and production teams in the books group, led by Lisa Thomas, as well as communications director Ann Day, marketing director Daneen Goodwin, and creative director Melissa Farris, who designed this book. It is a joy working with all of you. Katherine Shaw provided spot-on fact-checking, Jenna West helped with valuable research assistance in a pinch, and Allison Bruns crafted the brilliant illustrations that bring the people in these pages to life. I could not have made it to the finish line without the inestimable Anne Staub, who helped usher this book through its editorial stages with a perfect mix of rigor and understanding.

My fascination with the elements of genius—intelligence, creativity, perseverance, and luck—was prompted by a series of articles I wrote for *National Geographic* magazine: "Genius," about the qualities that spur great achievement (May 2017); "Intense Provocative

Acknowledgments

Disturbing Captivating Genius: Picasso," about Pablo Picasso's journey from prodigy to icon (May 2018); and "Leonardo's Enduring Brilliance," about Leonardo da Vinci's prescient vision and thirst for knowledge, tied to the 500th anniversary of his death (May 2019). The reporting I undertook for these stories underlies many of the themes woven into these pages; I drew from the Genius piece for the Introduction, and from the Picasso and Leonardo stories for Chapter 1 and the Epilogue. My deepest gratitude to the brilliant team at the magazine who supported and edited my writing: Susan Goldberg, David Brindley, Debra Adams Simmons, Jamie Shreeve, and Rachel Shea. Brainstorming sessions with photo editors Todd James and Kurt Mutchler enriched my work beyond measure. I am indebted to photographers Paolo Woods and Gabriele Galimberti for their artistry, inspiration, and camaraderie. *Grazie mille* for the best reporting trips ever.

Leonardo da Vinci captivated me from the start. I launched my reporting at an extraordinary conference, "Leonardo da Vinci: A Celebration," at the Aspen Institute in the summer of 2017. I extend my deep appreciation to host Walter Isaacson and the scholars I met there, along with many others who schooled me in the artist and the Renaissance, especially: Carmen Bambach, Martin Clayton, Paolo Galluzzi, Martin Kemp, Domenico Laurenza, Anne Leader, Stefania Marvogli, François Saint Bris, William Wallace, Francis Wells, and Sławomir Zubrzycki. Jesse Ausubel, Thomas Sakmar, and Karina Aberg, of the Leonardo da Vinci DNA Project, invited me into their world and provided indispensable insights that illuminated my understanding of the science.

The biographical material in this book would not exist without the work of historians and journalists who documented the lives of the 12 individuals profiled in these pages. My appreciation to them and to the many experts I queried, who shared their time and their vast knowledge. Although I could not quote all of them in these pages, they shaped my understanding of complex material, and their contributions were invaluable.

One of my greatest thrills was talking with family members who generously shared memories and insights: Claude Picasso, Bernard Ruiz-Picasso, Diana Widmaier Picasso, and Olivier Widmaier Picasso (all of whom I interviewed for my magazine feature and whose reflections I include in Chapter 1); Charlie Black and Susan Falaschi; Ellen Ford, John Blakely, Ford Blakely, and Jesse Itzler; David Roosevelt and Manuela Roosevelt; and Jane Kallir and Will Moses. Thank you for trusting me with your stories.

Family and friends are vital to sustaining any writerly enterprise. To those I named in my first book: I salute you all with the same enormous gratitude again. My Alexandria baseball family and friends and beloved Luxmanor, Amherst, and *Newsweek* clans offered lively conversation and welcome fellowship. A special tip of the hat to Susan Berfield and Steve Fainaru, my treasured go-to writing confidantes, and to Sonia Daccarett and Alex Bernstein for their warm friendship as well as much appreciated lodging during a reporting trip out west. Huge thanks, always, to Nancy Edson for her good cheer and unfailing support. An exceptional cadre of readers took the time to comment on chapters: Kevin Brown, Jennie Johns, Marina Kalb, Jane Kallir, Richard Kogan, Dan McGinn, and Anne Underwood. I cannot thank you enough.

This book is dedicated to my grandparents, whose memories are a blessing to all who followed. They were the ones who made their way for the rest of us. My strong and spirited grandmothers filled me with love, and although I did not know my grandfathers, I know their stories. I hold deep within me that my grandfather Max proudly carried my father's newspaper articles around in his pocket—even though he could not read them in English. I wish I could have shared my writing with him and with Bella, Sue, and Sam.

Acknowledgments

My parents, Phyllis and Bernard Kalb, are my mentors. My mother, a storyteller, listened with enthusiasm as I shared my reporting discoveries, and she fueled my desire to learn more. My father, ever the inquisitive reporter, asked, "Do they live up to the hype?" He kept me on my toes. Huge thanks, as always, to my extended family, with special appreciation to my London cousins, who provided brilliant company and a place to stay while reporting, and to Mady and Marvin Kalb for their ongoing encouragement and support. Abundant hugs and gratitude to my sisters, their husbands, and their children: Tanah, Hilmar, Max, Talia, and Camila; Marina, David, Eli, and Wolf; Sarinah, Jaron, Leo, Susannah, and Bella; and to Bobby Dean Phillips, who is there for all of us.

Above all, my deepest love and appreciation to Steve, Molly, and Noah. One of the greatest thrills of this project was the travel that Steve, Noah, and I did together. I will never forget standing in Isaac Newton's bedroom, sun streaming in, with the two of them by my side. Little did we know at the time that Newton's plague years would become our pandemic, during which much of the writing of this book took place. Steve and Noah made it easy even when it was hard. Steve provided never ending encouragement and enthusiasm. As always, he helped me unravel complexities and think big, offered wise editorial commentary, and bolstered my confidence when I needed it most. Noah cheered me on and became a student of the great minds in these pages. Nothing could make me happier.

My greatest wish is that Noah and his big sister, Molly, now all grown up and boldly making her way in the world, are forever nurtured, provided opportunities, and buoyed by others as they grow and discover. May their life journeys—and those of their peers and future generations of young people around the world—be filled with adventure, love, passion, sustenance, and joy.

Index

Spark

Index

Index